AF834797

Remote Detection and Maritime Pollution

Remote Detection and Maritime Pollution

Chemical Spill Studies

Edited by

Stéphane Le Floch
Frédéric Muttin

iSTE

WILEY

First published 2020 in Great Britain and the United States by ISTE Ltd and John Wiley & Sons, Inc.

Apart from any fair dealing for the purposes of research or private study, or criticism or review, as permitted under the Copyright, Designs and Patents Act 1988, this publication may only be reproduced, stored or transmitted, in any form or by any means, with the prior permission in writing of the publishers, or in the case of reprographic reproduction in accordance with the terms and licenses issued by the CLA. Enquiries concerning reproduction outside these terms should be sent to the publishers at the undermentioned address:

ISTE Ltd
27-37 St George's Road
London SW19 4EU
UK

www.iste.co.uk

John Wiley & Sons, Inc.
111 River Street
Hoboken, NJ 07030
USA

www.wiley.com

© ISTE Ltd 2020
The rights of Stéphane Le Floch and Frédéric Muttin to be identified as the authors of this work have been asserted by them in accordance with the Copyright, Designs and Patents Act 1988.

Library of Congress Control Number: 2020942928

British Library Cataloguing-in-Publication Data
A CIP record for this book is available from the British Library
ISBN 978-1-78630-639-5

Contents

Forewords

The challenges related to the protection and preservation of marine ecosystems rank among the top priorities of the French Ministry for the Ecological and Inclusive Transition. Over the past years, our involvement has led, at a national level, to the development of the French law on reclaiming biodiversity, which comprises several initiatives for the protection of the marine environment (recognition of the notion of ecological damage, creation of fishery conservation areas, requirement for ships traveling through the Pelagos and Agoa sanctuaries to be fitted with cetacean collision avoidance systems, supervision of the ballast water management of ships entering French waters, etc.).

At an international level, this involvement led to the selection of ocean protection as a theme for the debates held during COP 21 in Paris in 2015, and its inclusion in the Paris Agreement.

The ministry's commitment towards preserving the quality of our oceans is also illustrated through the support provided to CEDRE ever since its creation in 1979. This support is both financial and technical, in particular through regular, active participation in its different actions, as illustrated by the Information Day from which this book developed, which underscores CEDRE's position at the interface between science and the operational aspects inherent to spill response at sea.

Stéphanie CUBIER

Head of the Marine Environment Office at the
French Directorate of Water and Biodiversity
Ministry of the Environment, Energy and the Sea

Oil production at sea is one of the major uses of the ocean. To meet the technical challenges of this industry, it is our duty to preserve what is one of humanity's key resources. Total, as a leading light in responsible energy, examines and tests the latest advances geared towards minimizing the impact of its activities on the natural environment.

To maintain sufficient safety levels so as to prevent accidental releases, to determine the response actions to be rapidly implemented in the event of an incident and to more fully understand the behavior of oil at sea, Total depends on the research teams liable to support the scientific and technical developments identified so as to preserve the ocean and its ecosystems.

In order to maintain its remote sensing capabilities, Total deploys a research program that encompasses recent developments and promotes exchanges between the different players in the field.

True to its commitment towards meeting the environmental challenges and preserving the environment wherever the Group operates, Total implements anticipatory measures and environmental protection actions against oil pollution. This commitment is illustrated by the support it provides to the activities conducted by CEDRE, with which the Group enjoys close collaboration.

CEDRE's portfolio of activities clearly positions it at the crossroads of the needs expressed by the different stakeholders and of the proposed solutions promoted by research players.

The CEDRE Information Days, from which this book developed, are a chance to share, discuss and challenge these approaches, in order to pinpoint progress opportunities or even new research projects. Naturally, as an industrial player, Total both benefits from and greatly contributes to these debates.

Frédéric PÉRIÉ
President of CEDRE's Strategy Committee

Why was remote sensing chosen as the topic of this book? Such systems have been frequently used by CEDRE, in particular during the *Prestige* oil spill in 2002. Over and above operational aspects, it is also extremely useful for detecting deliberate discharge. In this context, strictly speaking, CEDRE does not conduct remote sensing, but rather works as a partner, in particular alongside the French Navy or within the framework of projects funded by the European Union. We are involved in equipment trials and tests, as well as in analyzing POLREPs, and we have been organizing aerial observation and remote sensing training courses for over 20 years. We also provide advice to the French and foreign authorities in charge of directing operations and of post-response legal aspects. The aim of this book is to offer as broad a vision as possible of remote sensing in the field of marine spills.

Stéphane DOLL
Director of CEDRE

PART 1

Remote Sensing Means

1

POLLUPROOF Project

1.1. Introduction

In cases of maritime pollution by HNS (Hazardous and Noxious Substances), specific methods of identification and characterization are needed. The project POLLUPROOF (which started in January 2014 and ended in mid-2017) aims to test and validate the use of optical sensing methods, including hyperspectral and radar sensors, in order to detect, locate and classify six HNS. In this chapter, the experimental approach followed during the project is detailed: the calibration of optical sensors in mesoscale experiments and the validation of optical and radar sensors in a realistic experiment at sea. The promising results obtained are specifically explained in the other chapters.

Maritime shipping activities are responsible for about 20% of the pollution at sea. Pollutants discharged accidentally or deliberately can endanger the biodiversity and eco-balance of our oceans. Exhaust emissions and cargo mishaps associated with an increase in vessel traffic are sources of pollution that affect both the marine environment (acidification, contamination of flora and fauna) and land (acid rain). This issue has become a priority at the national (Grenelle de la Mer) and regional (European – directives 2005/35 and 2005/33) levels, as demonstrated by the implementation of several international conventions (e.g. OPRC-HNS Protocol [OPR 00], MARPOL (completed in 1978) [MAR 73]). Obviously, the removal or drastic reduction of pollution resulting from maritime activities is a desirable objective. The

---

Chapter written by Sophie CHATAING, Sébastien ANGELLIAUME, Pierre-Yves FOUCHER, Eldon PUCKRIN and Stéphane LE FLOCH.

magnitude of the problem is highlighted by the quantity of goods transported by sea: of an estimated 8,000 million tonnes (Mt) of chemicals transported worldwide, 350 Mt are transported via European waterways. It is estimated that there are more than 100 incidents per year involving the illegal discharge of noxious liquid substances in these waters. For over 25 years, French Customs (DGDDI) have deployed aircraft equipped with remote sensing instruments (radar and scanner IR/UV), in order to successfully prosecute ships involved in oil spill incidents. The effectiveness of this policy has been demonstrated through a significant reduction in oil pollution in the waters under French jurisdiction (during the period between 2006 and 2012, the number of ships caught polluting was reduced by threefold).

This chapter presents the POLLUPROOF project through its objectives and the experimental approach used to achieve them. Results from the experimental parts are beyond the scope of this chapter and will be part of other chapters.

1.2. POLLUPROOF project

1.2.1. *Objectives*

The POLLUPROOF (PROOF improvement of HNS maritime POLLution by airborne radar and optical facilities) project would enhance the capabilities of French Customs to detect, locate and classify pollutants (other than hydrocarbons) originating from ship emissions (including particulates), in order to collect evidence for the prosecution of offenders while ensuring an effective intervention in the case of accidental discharge at sea.

The project is funded by ANR ECO-TECH 2013, and the members of the consortium have a recognized and complementary expertise in the field of aerial detection and marine pollution: ONERA, DGDDI, CEDRE, CEPPOL, Agenium, AVDEF and DRDC. In addition to the consortium, Transport Canada (TC) acts as an end-user and member of the steering committee. The project began in January 2014 and concluded in mid-2017.

The objectives of this project are:

1) to verify the ability to detect, locate and classify at least three of the six most noxious liquid substances transported by sea in Europe;

2) to achieve a reduction of spilled noxious liquid substances equivalent to the level for hydrocarbon emissions;

3) to develop a stronger policy to control the release of noxious gases within the sulfide emission control areas (SECA).

These objectives will be achieved by:

– deployment of radar (SAR/SLAR) and optical sensing (hyperspectral cameras) capabilities for detecting liquid pollutants at sea;

– evaluation of the complementarity of optical and radar information;

– identification of gaseous discharges of engine emissions and liquid pollutants using hyperspectral analysis.

To accomplish these activities, the POLLUPROOF project will analyze the needs of French Customs regarding aerial detection and will proceed with:

– calibration of optical measurements on liquid pollutants in mesoscale (test-tank) experiments located at CEDRE;

– airborne measurements of sea spills using hyperspectral optical and radar sensors, following the test-tank analysis;

– algorithm development for detection, location and classification of pollutants. The consortium will then produce a data gathering evidence methodology. French Customs staff will evaluate the effectiveness and applicability of these advances using a human–machine interface.

1.2.2. *Hazardous and noxious substances*

Six chemical substances have been chosen to evaluate the capability of remote sensing sensors: rapeseed oil, fatty acid methyl ester (FAME), toluene, heptane, xylene and methanol. These chemicals are among the most transported substances by maritime freight in Europe. Methanol and liquid chemicals represent 46% of the 165 million tonnes annually transported by chemical carriers, while vegetable oil accounts for 29% [OLA 09]. Some of these chemicals are classified as the most noxious substances in the IBC Code (IMO website), which provides an international standard for the safe carriage by sea of HNS in bulk. These chemicals have already been involved in accidents at sea, for example *Poona* sank in 1971 with 600 T of rapeseed oil, *Grape One* sank in 1993 with 3,000 T of xylene, *Cape Horn* carrying a cargo of 14,000 T of methanol was seriously damaged by an explosion in the port of Livorno in 2003 [CED 15, CUN 15]. Rapeseed oil and FAME are part of the vegetable oil family; toluene, heptane and xylene are petrochemical products; methanol is part of the family of alcohols and derivatives. Their main properties are described below.

Rapeseed oil: rapeseed or colza oil is a vegetable oil obtained from crushed colza seeds. At ambient pressure and temperature, rapeseed oil is a viscous liquid with a specific gravity of 0.910. Rapeseed oil is insoluble in water and does not evaporate (vapor pressure below 0.01 kPa at 25°C); these characteristics classify rapeseed oil as a floater F in the SEBC.

FAME: fatty acid methyl esters are biofuel directly added to conventional fuels such as diesel. At ambient pressure and temperature, they are a liquid with a specific gravity of 0.888. This product is virtually insoluble in water (solubility of 0.023 $mg.L^{-1}$ at 20°C) and has a relatively low evaporative potential (vapor pressure of 0.42 kPa at 25°C) making it a floater F in the SEBC.

Toluene: toluene, also named methylbenzene or phenylmethane, is an aromatic hydrocarbon that is commonly used as a chemical reagent or solvent, particularly in the industrial sector. Toluene is a liquid at ambient pressure and temperature and has a specific gravity of 0.867. Toluene is nearly insoluble in water (535 $mg.L^{-1}$ at 25°C) and tends to evaporate relatively easily (vapor pressure of 2.91 kPa at 20°C). Considering the SEBC classification, toluene is a floating and evaporating (FE) substance.

Heptane: heptane is the generic term to identify one of the nine isomers of C_7H_{16}, and is a saturated hydrocarbon of the linear alkane family. This is a constituent of fuel and is used as an extraction solvent, a synthesis intermediate in the chemical industry and as a solvent for glues, inks, rubbers and plastics. At ambient pressure and temperature, heptane is a volatile liquid (6–7.7 kPa at 20°C) and nearly insoluble in water (< 2 $mg.L^{-1}$ at 20°C). With a specific gravity of 0.710, heptane is lighter than water and floats. According to the SEBC classification, heptane is considered as an evaporator E.

Xylene: xylene or dimethylbenzene is a group of aromatic hydrocarbons with one methyl derivative on benzene. It is naturally present in oil and can be observed in (diesel) engine exhaust gases, either as a residual oil chemical or formed during incomplete combustion. Xylene is also produced from oil in the petrochemical industry and is one of the 30 most produced chemicals in the USA. It is used in the printing, rubber and leather industries mainly as a solvent. Xylene is an inflammable liquid with a pleasant fragrance. Chemical properties are similar from one isomer to another. Its specific gravity of 0.87 makes it float on water. Xylene is slightly soluble in water (solubility below 20 $mg.L^{-1}$ at 20°C) and is not likely to evaporate (vapor pressure of 0.89 kPa at 20°C). Due to these characteristics, xylene is considered as an FE (floater and evaporator) in the SEBC classification.

Methanol: methyl alcohol or methanol is the simplest alcohol with the chemical formula CH_3OH. At ambient temperature, this polar liquid is used as antifreeze (for coolant, for example), solvent or fuel (in aeromodeling, for example). Methanol is not present in large amounts in nature and is industrially produced. It is mainly used as the basic material for chemical synthesis of more complex chemical products. Nearly 40% of methanol is converted into formaldehyde, which is then transformed into plastics, synthetic resins, paints, explosives or fabrics. Methanol is a light liquid (specific gravity of 0.791), volatile (vapor pressure of 12.3 kPa at 20°C), miscible in water, inflammable and toxic with a characteristic odor. These properties enable the classification of methanol as a DE (Dissolving and Evaporating) substance.

1.3. Experimental approach

The experimental approach is divided into two parts: first, the calibration of optical sensors on liquid pollutants in mesoscale experiments; second, airborne measurements of sea spills using hyperspectral optical and radar sensors.

1.3.1. *Calibration of optical sensors*

The calibration of optical measurements on liquid pollutants was realized in mesoscale (test-tank) experiments located at CEDRE (Brest) in October 2014. Three configurations were tested, with each one having specific objectives:

– vertical configuration: comparison between clean water and water covered by a chemical;

– horizontal configuration: detection of the gas cloud produced by the evaporation of the chemical;

– tank: evaluation of the influence of the slick thickness.

1.3.1.1. *Vertical configuration*

Eight different HNS (benzene, toluene, xylene, diethyl ether (DEE), rapeseed oil, propanol, methanol and heptane) were released into seawater inside a floating aluminum frame that was installed in the CEDRE pool, as presented in Figure 1.1. For each product, different spill volumes were used from 60 mL up to 5 L, and the seawater was thoroughly cleaned after each spill.

(a) (b)

Figure 1.1. *(a) Aluminum frame installed in the CEDRE pool. (b) Aerial lift with the three hyperspectral cameras*

Three different hyperspectral imaging systems from 0.4 to 12 µm were used during this campaign: two reflective sensors (NEO HySpex cameras) from 0.4 to 1 µm (VNIR) and from 1 to 2.5 µm (SWIR) and a thermal longwave (LWIR) sensor from 8 to 12 µm (Telops Hypercam). Hyperspectral imaging cameras were deployed inside an aerial lift at a height of 12 m above the pool with a nadir-looking geometry and a Bomem MR300 spectroradiometer was placed next to the edge of the pool. The three sensors were pointed towards the expected center of the slick.

The aim was to evaluate how hyperspectral sensors can contribute to the detection of pollutants in a nadir-looking geometry.

1.3.1.2. *Horizontal configuration*

The sensors used for this configuration were as follows: two reflective sensors (NEO HySpex cameras) from 0.4 to 1 µm (VNIR) and from 1 to 2.5 µm (SWIR), an ASD Fieldspec camera from 0.4 to 2.5 µm, a thermal longwave (LWIR) sensor from 8 to 12 µm (Telops Hypercam), a Bomem MR300 spectroradiometer and a P-iCATSI (Polarized Improved Compact ATmospheric Sounding Interferometer).

Heptane, toluene, xylene, methanol, rapeseed oil, diethyl ether, silicone oil, unleaded gasoline and oil were released (volumes from 2 to 10 L). Diethyl ether and methanol were discharged directly in the basin, whereas the other products were released inside the floating aluminum frame.

Except for the interferometer, which was placed in an aerial lift several meters away from the basin, all the sensors were placed next to the edge of the pool.

1.3.1.3. *Tank*

The aim of this test was to qualify the impact of the thickness of the slick on the measured spectra. For this configuration, the Bomem MR300 spectroradiometer and the ASD Fieldspec camera were used. Heptane, xylene and rapeseed oil were released at the surface of a black-painted metallic barrel full of seawater.

The cloud coverage during the trial did not enable the realization of the initially planned measurements, and supplementary measures must be realized in the ONERA laboratory.

1.3.2. *Evaluation of radar, optical and hyperspectral sensors at sea*

1.3.2.1. *General presentation of the experiment*

The experimentation took place in May 2015 over the French coast in the Mediterranean Sea at two locations: 42°46.8'N/6°2.0'E for the first release and 42°45.5'N/5°48.5'E for the second and third releases. One cubic meter of each of the six chemical products (presented in section 1.2.2) was released at sea and imaged by radar and optical airborne sensors for their evaluation in real conditions.

HNS releases were performed from the salvage, rescue and oil spill response vessel *Ailette* of the French Navy under the direction of CEPPOL (Centre of Practical Expertise in Pollution Response) and CEDRE. Each chemical product was contained in a one cubic-meter tank, in HDPE (High Density PolyEthylene) for non-aggressive HNS (rapeseed oil, FAME and methanol) and in metal for reactive or corrosive HNS (xylene, heptane and toluene). Each tank was inserted in a metallic structure equipped with two 220 L floaters to ensure the floatability of the system and a lifting strap to manipulate the tanks with the onboard crane (Figure 1.2). The release of the HNS was performed from a dinghy by pulling a rope that activates the opening of the tank. Due to the difference in density between seawater and HNS, the chemicals spread at the sea surface.

In order to follow the drifts of the HNS slicks, two drifting buoys were implemented. Their GPS positions were transmitted by satellite every 15 minutes.

(a) (b)

Figure 1.2. *One cubic-meter tank. (a) HDPE. (b) Metal*

Three aircraft imaged the HNS slicks:

– a Falcon 20 from AVdef company equipped with the experimental pod SETHI, including the ONERA sensors: two radars (SAR) at X and L bands, full polarization (HH, HV, VH, VV), one SWIR hyperspectral camera (HYSPEX), one CamV2 camera (visible range);

– a POLMAR aircraft from DGDDI equipped with IR sensor, UV sensor and SLAR radar at X-band;

– a Cessna C-303 chartered by DRDC equipped with an IR hyperspectral camera (TELOPS).

On board the *Ailette*, two optical cameras were installed by DRDC: a MWIR hyperspectral camera and a Bomem MR304 LWIR and MWIR spectroradiometer.

In order to characterize the behavior of HNS once released at sea (evaporation, dissolution, emulsification, etc.), sampling of the slicks, the water column and the atmosphere were planned.

1.3.2.2. *Experimental releases*

HNS releases were planned to be performed two at a time, i.e. two HNS were nearly simultaneously released and imaged by the aircraft. The most dangerous HNS were first released (heptane and toluene), then methanol and xylene for the second release and, finally, FAME and rapeseed oil for the third release.

Date	Time (UTC)	Wind speed (m/s)	Wind direction (from-deg)	Wave height (m)	Wave direction (from-deg)
May 18, 2015	17:00	8	255	0.5	240
May 22, 2015	13:00	7	315	2	270
May 22, 2015	16:00	7	315	1.75	270

Table 1.1. *Environmental conditions*

The methodology used to release HNS is ideally suited to good weather conditions and, in particular, for a sea state compatible with the lifting operations. The first release was performed as planned on May 18 thanks to good weather conditions. Then, a strong wind episode (May 19–21, force 8, 35–40 knots) forced us to replan the last two releases on May 22. Sea and weather conditions during the experimentation are described in Table 1.1.

The first release took place on May 18, from 16:40 to 17:30 UTC. The sea was calm. Heptane and toluene were released from two semi-submersible 1 m^3 tanks at 16:30 and 16:35 UTC, respectively (Figure 1.3). Heptane and toluene were released at a fixed position (separated by about 100 m). No samplings were realized for this release due to the reactivity of these HNS. Two drifting buoys were deployed simultaneously with the releases.

The airborne sensors implemented for this release were:

– SETHI: radar (X + L quad-pol) and optical (SWIR) sensors;

– DGDDI: SLAR and IR/UV.

(a) (b)

Figure 1.3. *Releases of heptane and toluene*

The positions of the two drifting buoys and the tanks at their opening and their recovery are shown in Figure 1.4. Buoys and tank data show a WSW direction drift. Toluene and heptane are highly volatile chemicals, and their persistence at sea surface is limited (prediction of 30 minutes persistence with the software CHEMAP). However, drifting buoys are of interest during such experiments to obtain *in situ* information, which enables us to readjust the prediction models.

The second release took place on May 22, from 13:25 to 13:55 UTC. There was a heavy swell, restricting any activity at sea; hence, products were discharged directly from the back of the vessel (French Navy) advancing towards the east at a speed of 1 knot: methanol was released from 12:35 to 12:45 UTC and xylene from 12:55 to 13:25 UTC. The sea state made it impossible to use the dinghy, to deploy the drifting buoys or to sample the slick and water column.

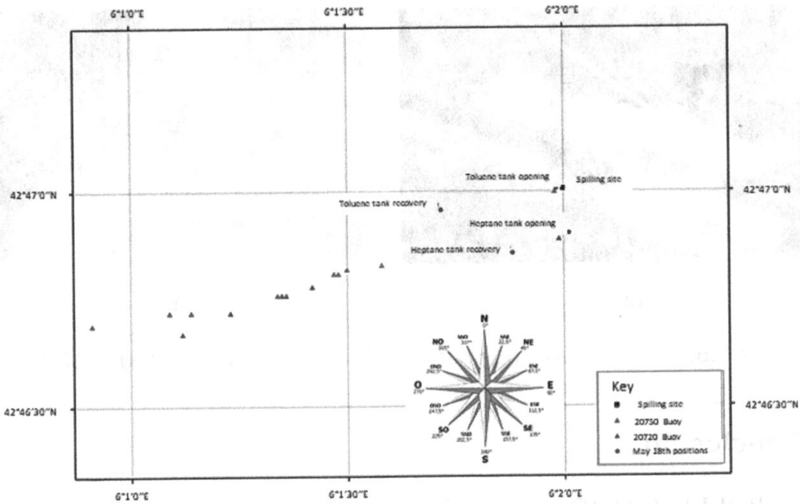

Figure 1.4. *Map of sea surface of tank and drifting buoy positions during the first release. For a color version of the figure, see www.iste.co.uk/lefloch/remote.zip*

The airborne sensors implemented for this release were:

– SETHI: radar (X + L quad-pol) and optical (SWIR) sensors;

– DGDDI: SLAR and IR/UV;

– DRDC: LWIR hyperspectral.

The third release took place on May 22, from 15:20 to 16:50 UTC. Because of the swell, products were also released directly from the back of the boat advancing towards the east at a speed of 1 knot: rapeseed oil was discharged from 15:00 to 15:30 UTC and FAME from 15:25 to 15:40 UTC (Figure 1.5). No samplings and no deployment of drifting buoys were possible due to the sea state.

The airborne sensors implemented for this release were:

– SETHI: radar (X + L quad-pol) and optical (SWIR) sensors;

– DGDDI: SLAR and IR/UV;

– DRDC: LWIR hyperspectral.

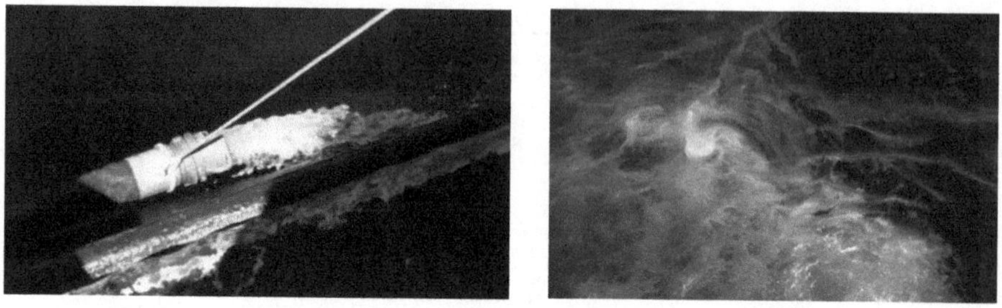

(a) (b)

Figure 1.5. *Releases from on board the* Ailette *(a) and rapeseed oil slick (b)*

1.4. Conclusion

The POLLUPROOF project aims to set up a procedure for collecting evidence of illegal maritime pollution from noxious liquid substances using airborne sensors. To achieve this goal, an experimental approach was divided into two parts. First, optical sensors were calibrated in a mesoscale spill experiment. Then, optical and radar sensors were implemented in a realistic experiment at sea to image the slicks from noxious substances.

The optical and radar results obtained during these experiments are beyond the scope of this presentation and will be presented elsewhere. However, they are clearly promising and demonstrate how hyperspectral sensors are complementary to classic optical and radar sensors for the detection of chemicals at the sea surface.

The research presented in this chapter is part of the POLLUPROOF research program (ANR-13-ECOT-007) funded by the French National Research Agency (ANR). The authors are very grateful to everyone involved in the experiment at sea (ONERA, CEDRE, AVDEF, DGDDI and the French Navy).

1.5. References

[CED 15] CEDRE ACCIDENTAL WATER POLLUTION, Database of spill incidents and threats in waters around the world, available at: http://wwz.cedre.fr/en/Resources/Spills, (accessed June 2020), 2015.

[CUN 15] CUNHA I., MOREIRA S., SANTOS M.M., "Review on hazardous and noxious substances (HNS) involved in marine spill incidents – An online database", *Journal of Hazardous Materials*, vol. 285, pp. 509–516, 2015.

[MAR 73] MARPOL, International convention for the prevention of pollution from ships, 1973, as modified by the Protocol of 1978 relating thereto and by the Protocol of 1997, International Maritime Organization (IMO), available at: http://www.imo.org (accessed June 2020), 1973.

[OLA 09] OLAFSEN G., "Chemical tanker trade", *Chemical and Product Tanker Conference, Back to Fundamentals*, London, UK, March 10–11, 2009.

[OPR 00] OPRC, Protocol on Preparedness, Response and Co-operation to pollution Incidents by Hazardous and Noxious Substances, International Maritime Organization (IMO), available at: http://www.imo.org, (accessed June 2020), 2000.

2

Multifrequency Radar Imagery and Characterization of Hazardous and Noxious Substances at Sea

2.1. Introduction

The increase in maritime traffic, and particularly the transport of chemical products, necessitates the enhancement of methods of prevention of and intervention in environmental chemical spills. Maritime pollution by chemical products occurs at a much lower frequency than spills of oil; however, the consequences of a chemical spill can be more wide-reaching than those of oil. While detection and characterization of hydrocarbons have been the subject of numerous studies, detection of other chemical products at sea using remote sensing has been little studied and is still an open subject of research. To address this knowledge gap, an experiment was conducted in May 2015 in the Mediterranean Sea in which controlled releases of hazardous and noxious substances were imaged by remote sensing systems. The aim of these experiments and subsequent analysis is to establish a procedure for collecting evidence of illegal maritime pollution by noxious liquid substances using airborne sensors.

In this chapter, we discuss the experimental procedure, which was carried out in collaboration with the French Navy and Customs, and report the main results from the airborne radar imaging campaign. We develop an accurate method for using multifrequency radar sensors to detect and quantify the

Chapter written by Sébastien ANGELLIAUME, Brent MINCHEW, Sophie CHATAING, Philippe MARTINEAU and Véronique MIEGEBIELLE.

impact of chemical products at sea. We conclude by demonstrating the capability of radar imagery to distinguish two different substances within the same spill.

Airborne and space-borne radar remote sensing is often used for oil slick detection over maritime surfaces [BRE 05, GIR 05, GAR 09, SOL 12]. In the oil and gas sector, oil slick detection using remote sensing is of great interest for exploration and environment program domains. In an operational context, oil slicks are usually detected using airborne and/or space-borne synthetic aperture radar (SAR). Indeed, the oil layer on top of the sea surface damps the so-called "capillarity waves", which are the origin of the sea surface roughness that significantly contributes to backscattered signal at a high frequency (microwave domain). Once detected, the spill is characterized using optical imagery [LEI 12].

Unlike hydrocarbons, there is limited research on the detectability of hazardous and noxious substances (HNS) at sea using remote sensing. In that context, an experimental campaign of measurements (called POLLUPROOF) was conducted in May 2015 in the Mediterranean Sea (off the French coast). Controlled release of six chemical and non-hydrocarbon oil products was carried out in collaboration with the French Navy and Customs. Polarimetric SAR (POLSAR) data were acquired at X- (9.75 GHz) and L-band (1.325 GHz) simultaneously by SETHI, the ONERA airborne SAR system [BON 09], for released products. Hyperspectral data (SWIR and LWIR) have also been acquired by ONERA and RDDC; French customs have also participated in the exercise at sea with their airborne surveillance system (X-band RAR, IR and UV imagery), but RAR (Real Aperture Radar) and optical remote sensing is beyond the scope of this chapter.

This chapter first examines the capability of high-resolution multifrequency SAR imagery to detect HNS at sea and then the potential of radar imagery to quantify and characterize the spill: thickness or amount of product and distinction between different products.

This chapter is organized as follows. Section 2.2 describes the experimental measurement campaign at sea. Section 2.3 presents the methodology for detecting and characterizing HNS. Finally, section 2.4 presents some original results and demonstrates the relevance of radar imagery for monitoring non-hydrocarbon pollution over maritime surface.

2.2. Experimentation at sea

2.2.1. *Radar imagery*

SETHI is the airborne remote sensing imaging system developed by ONERA [BON 09]. It integrates a new generation of radar and optronic payloads. It can operate over a wide range of frequency bands from UHF–VHF to X-band, including L-band with long range, very high resolution and polarimetric and interferometric capabilities. SETHI is a pod-based system operating onboard a Falcon 20 Dassault aircraft, which is the property of the AvDEF company.

For the POLLUPROOF campaign, quad-pol SAR data have been acquired simultaneously at X- and L-bands, with resolutions of 0.5 and 1.0 m, respectively. Images are generated with an azimuth resolution equal to the range resolution at both X- and L-bands, which implies an integration time equal to 1.1 s at X-band and 4.1 s at L-band. The main characteristics of acquisition are summarized in Table 2.1.

Frequency	Bandwidth	Polarization	Incidence angle	Swath
X	300 MHz (9.6–9.9 GHz)	Quad-pol (HH, HV, VH, VV)	45° (34–52°)	1,500 m (slant range)
L	150 MHz (1.25–1.4 GHz)	Quad-pol (HH, HV, VH, VV)	45° (34–52°)	1,500 m (slant range)

Table 2.1. *SETHI – waveform*

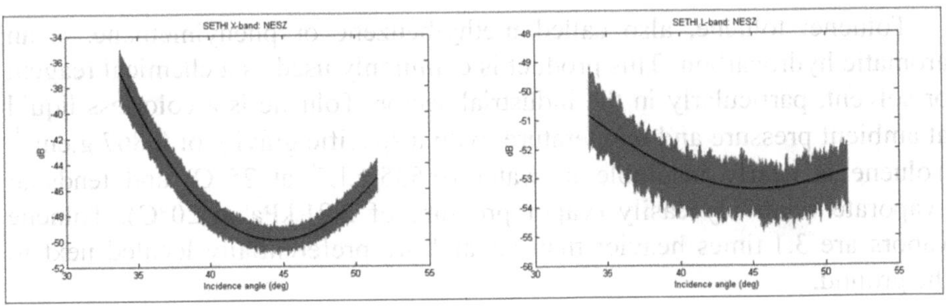

Figure 2.1. *SETHI – instrumental noise at X-band (left) and L-band (right)*

Incidence angle varies across the swath from 34° to 52°. Instrumental noise floor has been estimated using the method proposed by Hajnsek *et al.* [HAJ 03], and the results are shown in Figure 2.1. The noise equivalent sigma0 (NESZ) is very low, allowing a sufficient signal-to-noise ratio over the spill for efficient analysis.

2.2.2. *Chemical products*

Six chemical substances have been chosen to evaluate the capability of remote sensing sensors. The choice was made to cover different chemical families in order to be as representative as possible of chemical products often transported by sea and classified as noxious substances. The chemical substances selected are the following:

– category I: vegetal oil and fatty acid esters:

- rapeseed/colza oil,

- fatty acid methyl esters (FAME),

– category II: petrochemical products:

- toluene,

- heptane,

- xylene,

– category III: alcohols and derivatives:

- methanol.

Toluene: toluene, also called methylbenzene or phenylmethane, is an aromatic hydrocarbon. This product is commonly used as a chemical reagent or solvent, particularly in the industrial sector. Toluene is a colorless liquid at ambient pressure and temperature, with a specific gravity of 0.867 $g.cm^{-3}$. Toluene is nearly insoluble in water (0.535 $g.L^{-1}$ at 25°C) and tends to evaporate relatively easily (vapor pressure of 2.91 kPa at 20°C). Toluene vapors are 3.1 times heavier than air and are preferentially located next to the ground.

Heptane: heptane is the generic term to identify one of the nine isomers of C_7H_{16} and is a saturated hydrocarbon of the linear alkane family. This is a constituent of fuel and is used as an extraction solvent, a synthesis intermediate in the chemical industry and a solvent for glues, inks, rubbers and plastics. At ambient pressure and temperature, heptane is a colorless liquid, volatile (6–7.7 kPa at 20°C) and nearly insoluble in water (< 2 $mg.L^{-1}$). With a specific gravity of 0.710 $g.cm^{-3}$, heptane is lighter than water and floats.

FAME: fatty acid methyl esters are biofuel directly added in conventional fuels such as diesel. At ambient pressure and temperature, they are a liquid with a specific gravity of 0.888 $g.cm^{-3}$. This product is highly insoluble in water (solubility of 0.023 $mg.L^{-1}$ at 20°C) and practically does not evaporate (vapor pressure of 0.42 kPa at 25°C).

Methanol: methyl alcohol or methanol is the simplest alcohol with chemical formula CH_3OH. It is a colorless light liquid (specific gravity of 0.791 $g.cm^{-3}$), volatile (vapor pressure of 12.3 kPa at 20°C), miscible in water, inflammable and toxic with a characteristic odor. At ambient temperature, this polar liquid is used as an antifreeze (for coolant, for example), solvent or fuel (in aeromodeling, for example). Methanol is not present in large amounts in nature and is industrially produced. Methanol is mainly used as a basic material for the chemical synthesis of more complex chemical products. Nearly 40% of methanol is converted into formaldehyde to be then transformed into plastics, synthetic resins, paints, explosives or fabrics.

Rapeseed oil: rapeseed or colza oil is a vegetal oil obtained by crushing colza seeds. This is the second most consumed food oil in France after sunflower oil. At ambient pressure and temperature, rapeseed oil is a viscous yellowish liquid with a specific gravity of 0.910 $g.cm^{-3}$. Rapeseed oil is insoluble in water and does not evaporate (vapor pressure below 0.01 kPa at 25°C).

Xylene: xylene or dimethylbenzene is a group of aromatic hydrocarbons with one methyl derivative on benzene. It is naturally present in oil; xylene is observed in (diesel) engine exhaust gases, either as a residual oil chemical or formed during incomplete combustion. Xylene is also produced from oil in the petrochemical industry and is one of the 30 most produced chemicals in the United States. It is used in the printing industry, rubber and leather industries mainly as a solvent. Xylene is a colorless and very inflammable liquid with a pleasant fragrance. The chemical properties of its isomers are similar. Its specific gravity of 0.87 $g.cm^{-3}$ means it floats on water. Xylene is slightly soluble in water (solubility below 20 $mg.L^{-1}$) and is not likely to evaporate (vapor pressure of 0.89 kPa at 20°C). Xylene vapors are heavier than air and tend to be located near the ground.

These chemicals are some of the most transported substances by maritime transport in Europe, and some of them are classified as the most noxious substances in the IBC Code (IMO) [INT 11]. During the POLLUPROOF experiment, 1 m^3 of each of these six products was released at sea and imaged by airborne remote sensing sensors.

2.2.3. *Planning of measurements*

Experimentation took place in May 2015. Three flights were performed, with two releases per flight:

– the first flight took place on May 18, from 16:40 to 17:30 UTC. The sea was calm. Heptane and toluene were released from two semi-submersible 1 m^3 tanks at 16:30 and 16:35 UTC, respectively (Figure 2.2). Heptane and toluene were released at a fixed position (separated by about 100 m); SAR acquisitions began 10 min after the first release (heptane) and 5 min after the second release (toluene);

– the second flight took place on May 22, from 13:25 to 13:55 UTC. There was a heavy swell, restricting any activity at sea; hence, products were released directly from the back of the boat (French Navy) advancing towards the east at a speed of 1 knot: methanol was released from 12:35 to 12:45 UTC, and xylene from 12:55 to 13:25 UTC (Figure 2.3). SAR acquisitions began 40 min after the end of the first release (methanol) and at the end of the second release (xylene);

– the third flight took place on May 22, from 15:20 to 16:50 UTC. Because of the swell, products were also released directly from the back of the boat advancing towards the east at a speed of 1 knot: rapeseed oil was released from 15:00 to 15:30 UTC, and FAME from 15:25 to 15:40 UTC (Figure 2.4). SAR acquisitions were synchronized with the release.

Sea and weather conditions during experimentation are described in Table 2.2.

Date	Time (UTC)	Wind (from)		Wave (from)	
		Speed (m/s)	Direction (deg)	Height (m)	Direction (deg)
May 18, 2015	17:00	8	255	0.5	240
May 22, 2015	13:00	7	315	2	270
May 22, 2015	16:00	7	315	1.75	270

Table 2.2. *Environmental conditions during experimentation at sea*

Figure 2.2. *First flight: heptane and toluene release from semi-submersible tank*

Figure 2.3. *Second flight: methanol and xylene release from the boat*

Figure 2.4. *Third flight: rapeseed oil (left) and FAME (right) release from the boat*

2.3. Methodology

2.3.1. *Scattering from ocean surface*

For the frequency bands mainly used in Earth observation (X-, C- and L-band) and for incidence angles ranging from 30° to 60°, an ocean surface is a randomly rough surface where the radar backscatter is dominated by the Bragg scattering mechanism. As a result, the radar backscattered power, which is commonly defined by the normalized radar cross-section (NRCS), is greater in VV polarization than in HH and HV [VAL 78]. For each polarization, the NRCS is proportional to the spectral energy density of the sea surface waves with wavelength (λ_{sea}) that satisfies:

$$\lambda_{sea} = \frac{\lambda_{EM}}{2\sin(\theta_i)} \qquad [2.1]$$

where λ_{EM} and θ_i are the wavelength and the local incidence angle of the electromagnetic (EM) waves transmitted by the radar system, respectively. Ocean wavelengths corresponding to the Bragg wavelength are shown in Figure 2.5 for the three most common frequency bands (X-, C- and L-bands). It is obvious that the carrier frequency has a crucial impact and using different data acquired at different frequencies over the same area can provide additional information and improve our ability to characterize the imaged surface. While X- and C-bands are relatively close, we can assume that the use of X- and L-bands will allow us to obtain simultaneous information on different scales of the sea.

Ocean surface is usually modeled as a composition of slightly rough tilted facets, each of which has superimposed small-scale surface roughness that creates a Bragg scatterer. Small-scale roughness is randomly distributed on the scattering surface and responds to the strength of local wind (i.e. gravity capillarity waves). The tilt of the facet is caused by larger-scale gravity waves on the ocean surface. The orientation of the facet normal in the radar reference frame is defined by two angles ψ and ξ. The resulting local incidence angle of the EM wave is:

$$\theta_i = cos^{-1}[\cos(\theta + \psi) \cos \xi] \qquad [2.2]$$

where θ is the EM angle of incidence relative to local, un-tilted up.

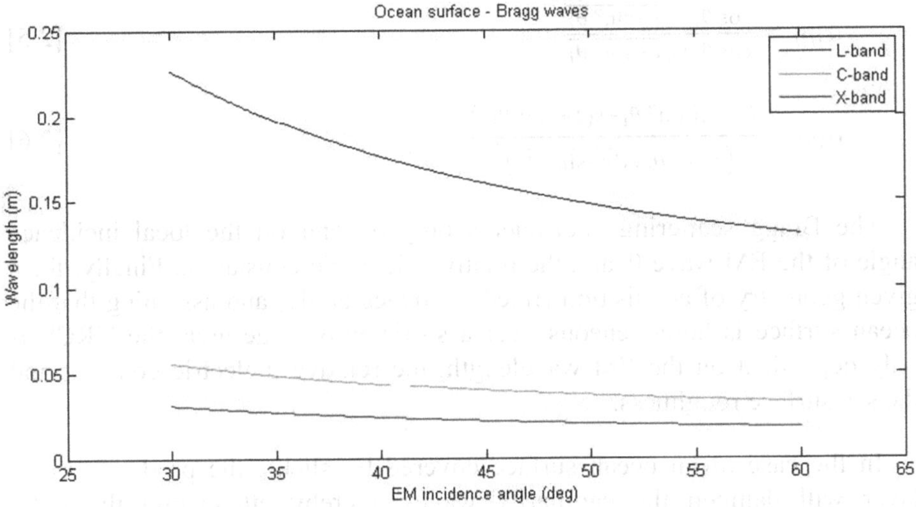

Figure 2.5. *Ocean waves in resonance with the EM signal (Bragg mechanism). For a color version of the figure, see www.iste.co.uk/lefloch/remote.zip*

Following this approach, the sea may be modeled as the superposition of two independent processes describing the small- and large-scale components. The two-scales model (TSM), based on a spectral description of the sea surface, has been proposed to formulate this composite-surface scattering [ELF 97, SOR 08]. The NRCS is then given by [VAL 78]:

$$\sigma_{pp}^{0} = 4\pi\kappa_{EM}^{4} \cos^{4}\theta_{1}\Gamma_{pp}W \qquad [2.3]$$

where the subscript p denotes either H or V polarization, $k_{EM} = 2\pi/\lambda_{EM}$ is the EM wavenumber, W is the spectral density of the ocean surface roughness and Γ_{pp} is the reflectivity defined by:

$$\Gamma_{pp} = \left|\left(\frac{\sin(\theta+\psi)\cos\xi}{\sin\theta_i}\right)^2 \alpha_{pp} + \left(\frac{\sin\xi}{\sin\theta_i}\right)^2 \alpha_{qq}\right|^2 \qquad [2.4]$$

where the subscript q (q ≠ p) denotes either H or V polarization. The reflectivity depends on the facet tilt, the EM wave incidence angle and the scattering coefficients such that [VAL 78]:

$$\alpha_{HH} = \frac{\cos\theta_i - \sqrt{\varepsilon - \sin^2\theta_i}}{\cos\theta_i + \sqrt{\varepsilon - \sin^2\theta_i}} \quad [2.5]$$

$$\alpha_{VV} = \frac{(\varepsilon - 1)\left(\sin^2\theta_i - \varepsilon\left(1 + \sin^2\theta_i\right)\right)}{\left(\varepsilon\cos\theta_i + \sqrt{\varepsilon - \sin^2\theta_i}\right)^2} \quad [2.6]$$

The Bragg scattering coefficients only depend on the local incidence angle of the EM wave θ_i and the relative dielectric constant ε. Finally, for a given geometry of acquisition (fixed incidence angle) and assuming that the ocean surface is homogeneous over a sufficiently large area, the NRCS is only dependent on the EM wavelength, the relative dielectric constant and the sea surface roughness.

In the case of an ocean surface covered by slicks, the product surface layer will dampen the capillarity waves, thereby attenuating the radar backscattered power by diminishing the surface roughness. The relative dielectric constant can also be modified in the case of a product which mixes with sea water. The resulting dielectric constant will be smaller than that for pure sea water, because the relative dielectric constant of HNS is low compared to that of sea water [MIN 12a]. The resulting NRCS will be lower for a product that mixes with water than for clean sea surface.

The starting point of this study is based on the assumption that in the case of a thin layer, only short capillarity waves will be damped; when the layer gets thicker, longer capillarity waves will also be damped. These phenomena should be observable in the case of multifrequency SAR data acquired simultaneously over the same surface. Moreover, this should be enhanced when using EM wavelengths sufficiently apart (e.g. X- and L-band radar frequencies) since the EM signal will simultaneously interact with hydrodynamic mechanisms of different scales (Figure 2.5).

2.3.2. *Detection and relative quantification*

Many studies have suggested using polarimetric SAR (POLSAR) parameters to improve slicks detection capability (see the review by Solberg) [SOL 12]. A comprehensive comparison of those parameters was undertaken by Angelliaume *et al.* [ANG 15]. Kudryavtsev *et al.* and Angelliaume *et al.* [KUD 12, ANG 15] quantitatively demonstrate the effectiveness of the polarization difference (PD = VV − HH in linear units) for slick detection on the ocean surface. The utility of this parameter for a marine pollutant detection application is that polarization difference (PD) is essentially proportional to the wave number spectrum taken at the Bragg frequency [GUE 10]. It is therefore very sensitive to small-scale features of the ocean surface (damped by the slick) and quite insensitive to larger scales, which rather drive the unpolarized contribution to the NRCS.

The main drawback of the PD is that it is not normalized, making it difficult to define a threshold with which a detection map can be established. To overcome this limitation, we propose a normalized variant of PD. We note that PD ranges from a maximum value (PD_{max}) that occurs in the case of a clean sea surface and diminishes to 0 as the impact of the substance on the surface increases, since both VV and HH reflectivities tend to 0 over a perfectly smooth surface. Hence, we define the normalized polarization difference (NPD) as:

$$NPD = \frac{PD_{max} - PD}{PD_{max}} \quad 0 \leq NPD \leq 1 \qquad [2.7]$$

NPD is equal to 0 in the case of a clean sea surface and increases to 1 as the impact of the substance on the ocean surface increases. PD_{max} is the polarization difference value in the case of a clean sea surface and can be estimated through a physical two-scale model [SOR 08, ANG 14] or an empirical model [GRE 12], using wind speed and wind direction information.

NPD can be interpreted as an indication of the presence and the impact of a substance on the ocean surface. It can be used directly or thresholded to produce a binary detection map.

2.3.3. *Oil/water mixing index*

The basic premise of the oil/water mixing index (M_{dex}) is that radar backscattered power is diminished by oil slicks through mechanical damping of Bragg-wavelength capillary waves and reductions in the relative dielectric constant of the upper few centimeters near the sea surface [MIN 12b, MIN 12a]. By decoupling the relative contribution to signal attenuation of mechanical damping of surface waves and changes in dielectric constant, we can define the characteristics of the slick along a spectrum ranging from thin surface films to thicker emulsions. This information is critical for efficient clean-up operations.

In addition to the assumptions underlying the NRCS model (equation [2.3]), the process of decoupling the mechanical and electromagnetic attenuation mechanisms requires us to assume that: (1) the long-wavelength tilting of the sea surface (as described by angles ψ and ζ in equation [2.2]) is unaffected by the presence of an oil slick; and (2) the dielectric constant of uncontaminated seawater is well constrained. These assumptions allow for inferences of the long-wavelength tilting of the sea surface over the entire radar image (so long as the radar images ample areas of uncontaminated water), the dielectric constant of contaminated water and the small-scale roughness spectrum of both contaminated and uncontaminated water. As described by Minchew [MIN 12a], the salient parts of the process of decoupling the mechanical and electromagnetic attenuation mechanisms can be summarized as follows:

1) use the co-polarized ratio over uncontaminated seawater and an assumed value for the dielectric constant of pure seawater, ε_r^{water}, to infer the long-wavelength tilting of the ocean surface;

2) calculate the short-wavelength roughness spectrum over uncontaminated water, W^{water}, by applying the tilt angles inferred in step 1;

3) for each pixel of contaminated water, infer the local (effective) relative dielectric constant, ε_r^{oil}, from the co-polarized ratio and the inferred long-wavelength tilt angles deduced from step 1;

4) calculate the short-wavelength roughness spectrum over contaminated water, W^{oil}, using the dielectric constant inferred in step 3 and the tilt angles obtained in step 1.

M_{dex} values can then be calculated as follows:

$$M_{dex} = M_W - M_\alpha \qquad [2.8]$$

$$M_W = \frac{W^{water} - W^{oil}}{W^{water}} \qquad [2.9]$$

$$M_\alpha = \frac{|\alpha_{VV}^{water}|^2 - |\alpha_{VV}^{oil}|^2}{|\alpha_{VV}^{water}|^2} \qquad [2.10]$$

where M_W is the normalized damping factor and M_α is the normalized power attenuation factor, both ranging from 0 to 1. M_W is a measure of how much the product will attenuate the sea surface roughness (0 = no damping; 1 = total damping); M_α is a measure of how much the backscattered signal is attenuated due to a decrease of the relative dielectric constant (0 = no attenuation; 1 = total attenuation).

M_{dex} ranges from −1 to 1. Negative values indicate that the decrease of the EM signal due to the presence of HNS on the ocean is more due to a decrease of the relative dielectric constant than to a decrease of the surface roughness, meaning that the product is mixed with the sea water. On the contrary, positive values indicate that the decrease of the EM signal is mainly due to a smoothing of the ocean surface and thus we observe the presence of a product that forms a film (more or less homogeneous) on the top of the sea surface. In an operational context, this information may be valuable to identify the behavior of the HNS and guide actions.

2.4. Results and discussion

2.4.1. *Observation of hazardous and noxious substances at sea*

Heptane and toluene: neither substance was observed in SAR images, at X- and L-bands, most likely because they are extremely volatile products, which do not affect the surface roughness. Remember that products (1 m^3 each) were released only 5 and 10 min before acquisitions.

Methanol and xylene: as methanol is an extremely volatile substance soluble in the water column and as SAR acquisition began 40 min after the end of the release, methanol was not observed in SAR images. On the other hand, the impact of xylene is clearly observable on X- and L-band SAR imagery, without obvious influence of the flight trajectory relative to the wind. Figure 2.6 shows illustrations of X-band VV polarization images acquired over xylene for upwind and crosswind observation. The size of the spill is about 1,750 m × 200 m; the bright point just at the end of the spill is the ship (53 m length) from which releases were performed.

Rapeseed oil and FAME: FAME and rapeseed oil are two persistent substances that were both clearly observable on SAR images acquired at X- and L-bands during the third flight. Figure 2.7 shows an illustration of

VV polarized images acquired simultaneously at X- and L-bands over the spill at 16:07 UTC. Rapeseed oil, having been released first, corresponds to the right part of the spill; FAME is on the left. From *in situ* information, we know that FAME ranges from azimuth 4,100 m to 5,500 m and that rapeseed oil ranges from azimuth 6,000 m to 8,500 m, and between we have a mixture of the two products.

While it is not possible to visually distinguish the FAME and rapeseed oil, we observe a significant difference between L- and X-band acquisitions. Contrast between the spill and the clean sea surface is more significant at X-band than at L-band. Figure 2.8 shows an azimuth profile calculated by averaging from 3,800 m to 3,900 m in range (the spill is between azimuth 4,800 to 5,600 m and azimuth 6,500 to 8,200 m). While at X-band the spill seems homogeneous (Figure 2.7 top), at L-band (Figure 2.7 bottom), we observe strong variation of the EM signal into the spill with dark patches, due to a stronger impact of the HNS on the sea surface roughness at wavelengths corresponding to the L-band Bragg wavelength. This observation is particularly important: as previously demonstrated by Wismann *et al.* [WIS 98] and Gade *et al.* [GAD 98], it is confirmed that the EM signal backscattered by HNS is dependent on the wavelength and using different frequency bands should allow us to better characterize the spill.

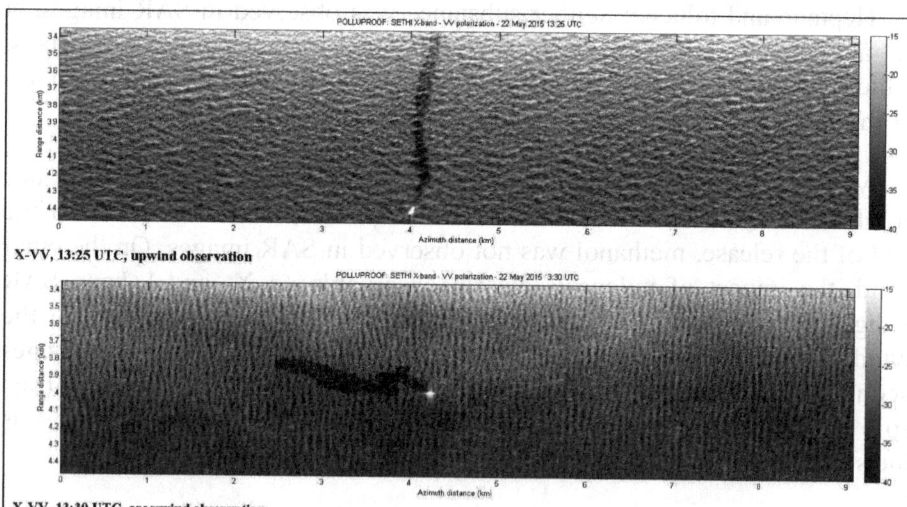

Figure 2.6. *X-VV SAR images over xylene – upwind 13:25 UTC (top) and crosswind 13:30 UTC (bottom)*

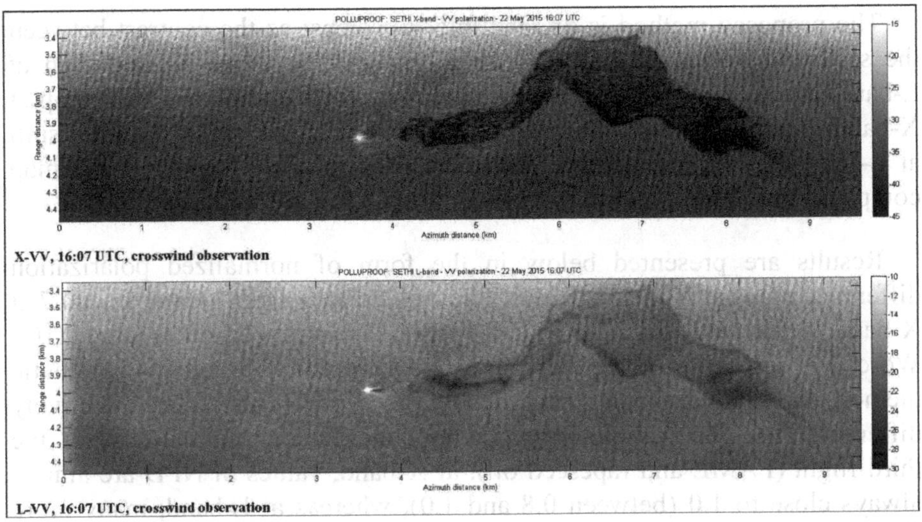

Figure 2.7. *SAR images over rapeseed oil and FAME – X-VV (top) and L-VV (bottom) – 16:07 UTC*

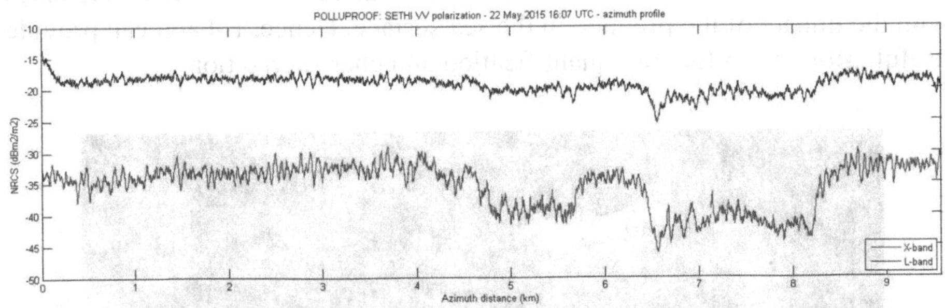

Figure 2.8. *Range profile at X-band (red) and L-band (blue). For a color version of the figure, see www.iste.co.uk/lefloch/remote.zip*

2.4.2. *Detection and quantification of impact on the ocean surface*

As presented in section 2.3.2, the normalized polarization difference (NPD) can be used for the detection of HNS at sea and quantification of their impact on the ocean surface.

The proposed method is divided into two steps: as the contrast between the spill and ocean clean surface is more pronounced at X-band than at L-band, a detection mask is first calculated by thresholding the NPD map at X-band. Then, since we observe more variation of the signal into the spill at L-band than at X-band, the detection mask is applied to the NPD map computed at L-band.

Results are presented below in the form of normalized polarization difference maps at X- and L-bands, thresholded for values greater than 0.7 at X-band, for xylene (Figure 2.9) and FAME and rapeseed oil (Figure 2.10). We can observe that, first, the spill is well identified using this automatic method, and second, the data provided by the two frequencies are effectively different. This observation is particularly conspicuous on the spill of the third flight (FAME and rapeseed oil): at X-band, values of NPD are almost always close to 1.0 (between 0.8 and 1.0), whereas at L-band, many more fluctuations are observed. At X-band, the response is saturated by the presence of the spill. This frequency is perfectly suited for detection but does not seem to provide information on the thickness. At lower frequency (L-band in our case), the EM signal backscattered by the slick fluctuates with the impact of the product on the sea surface. Hence, L-band can provide useful information for spill quantification and characterization.

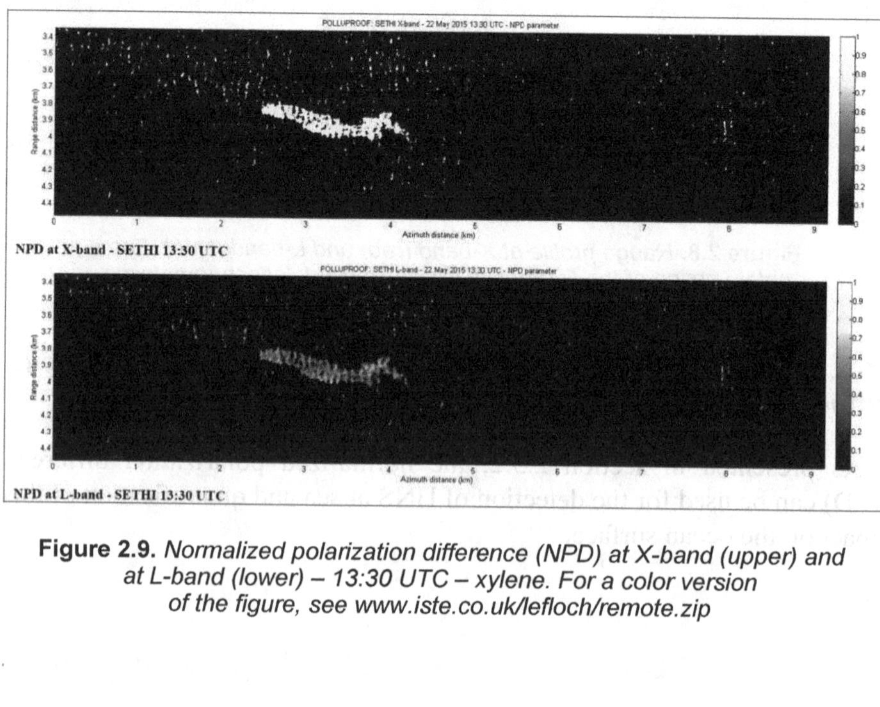

Figure 2.9. *Normalized polarization difference (NPD) at X-band (upper) and at L-band (lower) – 13:30 UTC – xylene. For a color version of the figure, see www.iste.co.uk/lefloch/remote.zip*

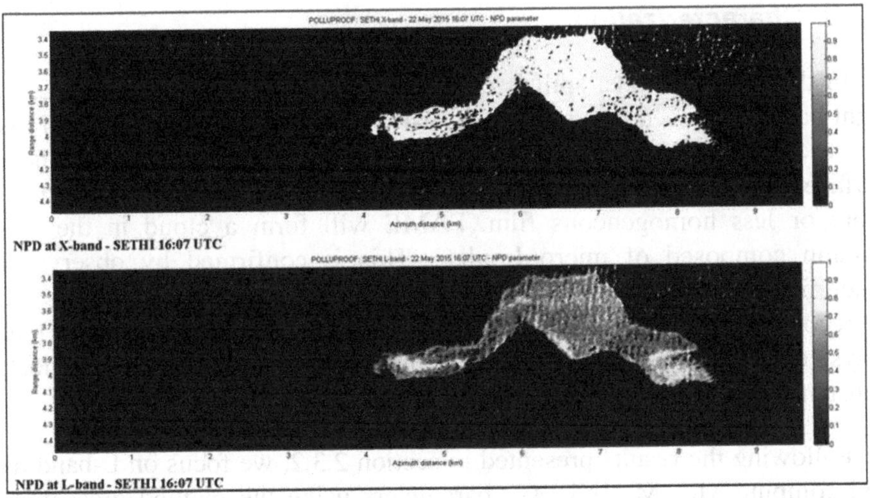

Figure 2.10. *Normalized polarization difference (NPD) at X-band (upper) and L-band (lower) – 16:07 UTC – FAME and rapeseed oil. For a color version of the figure, see www.iste.co.uk/lefloch/remote.zip*

Figure 2.11 shows two range profiles over xylene and FAME at X- and L-bands. Clearly, NPD values at X-band always range from 0.8 to 1.0, while the dynamics at L-band are greater. For FAME, the NPD value at L-band reveals a stronger impact of the HNS on the sea surface in the center of the spill than at the extremities. Information provided by normalized polarization difference and simultaneous use of two frequency bands allow us to detect HNS at sea and to quantify their impact on the ocean surface in terms of roughness. However, results shown in Figure 2.10 do not enable us to distinguish between the two products that form the spill (rapeseed oil and FAME).

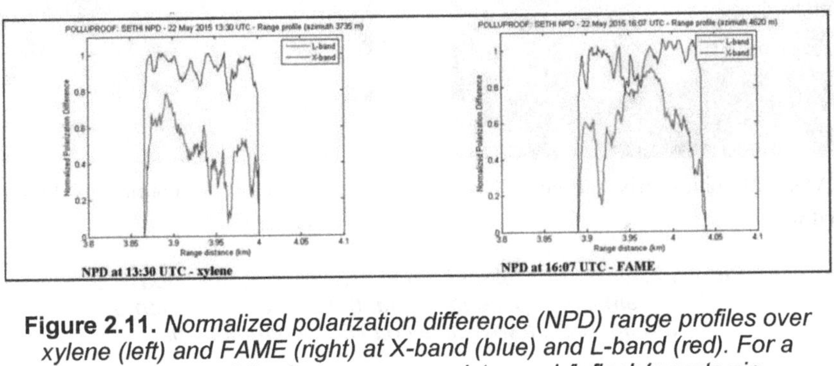

Figure 2.11. *Normalized polarization difference (NPD) range profiles over xylene (left) and FAME (right) at X-band (blue) and L-band (red). For a color version of the figure, see www.iste.co.uk/lefloch/remote.zip*

2.4.3. *Characterization*

In the third exercise, spill is composed of rapeseed oil and FAME. The right part is rapeseed oil, and the left part is FAME; in the middle, there is a mixture of two. We can expect different behaviors of each product on the sea surface: rapeseed oil is supposed to remain above the surface and produce a more or less homogeneous film. FAME will form a cloud in the water column composed of micro-droplets. This is confirmed by observations made from the ship during the releases (Figure 2.12). These behaviors must be recovered by SAR imagery as they impact the ocean surface in different ways: damping of capillarity waves and/or modification of the dielectric constant by mixing with sea water.

Following the results presented in section 2.3.2, we focus on L-band data and compute M_W, M_α and M_{dex} parameters using the method presented in section 2.2.3 and initially published in Minchew [MIN 12a]. As for the normalized polarization difference map, a detection map was used beforehand and values have been computed for each pixel detected as HNS using X-band data.

FAME: 3D plume/mix between HNS and water

Rapeseed oil: discontinuous film on the surface

Figure 2.12. *Photos taken from the ship during the release of FAME (left) and rapeseed oil (right). For a color version of the figure, see www.iste.co.uk/lefloch/remote.zip*

Figure 2.13 shows maps obtained over the full region of interest: the normalized damping factor (M_W), the normalized power attenuation factor (M_α) and the mixing index (M_{dex}). We can clearly observe the separation between the two products constituting the spill (see M_{dex}, Figure 2.13).

Figure 2.14 shows histograms (NPD at X- and L-bands, M_W, M_α and M_{dex}) over xylene, rapeseed oil and FAME. Xylene and rapeseed oil have similar responses: M_α is close to 0, M_{dex} is greater than 0 and M_W and NPD at L-band are equal, which corresponds to the presence of a film on the sea surface that damped the short waves. FAME has a different behavior: M_α is no longer negligible and thus M_{dex} is negative, which means that, as expected, mixing is more present over FAME than over rapeseed oil and xylene. For the three products, NPD at X-band is always significantly high (greater than 0.9). Over FAME, NPD at L-band allows us to distinguish two areas corresponding to two different concentrations, as can be seen in Figure 2.10.

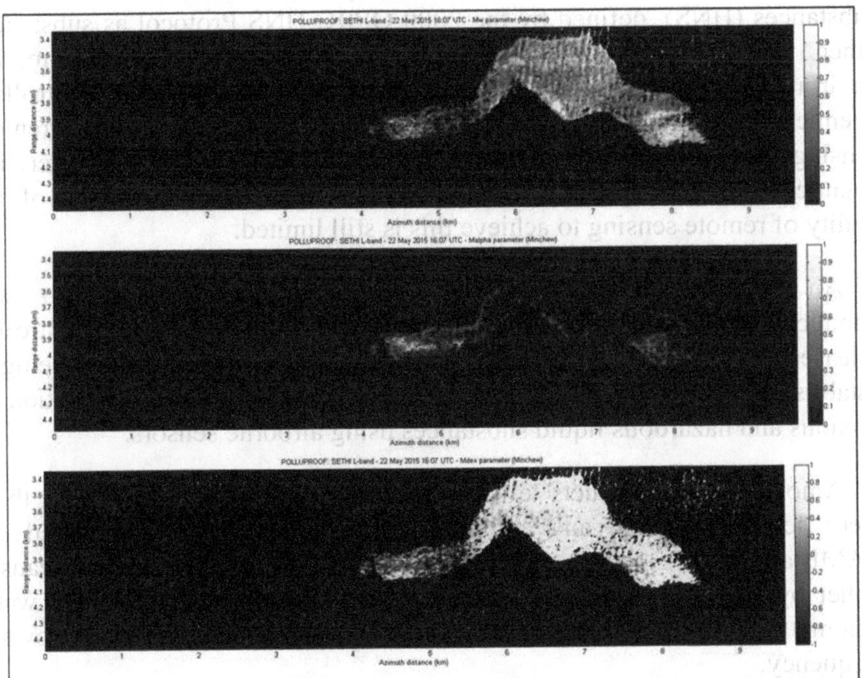

Figure 2.13. *MW (top), Mα (middle) and Mdex (bottom) – L-band SAR acquisitions at 16:07 UTC – FAME and rapeseed oil. For a color version of the figure, see www.iste.co.uk/lefloch/remote.zip*

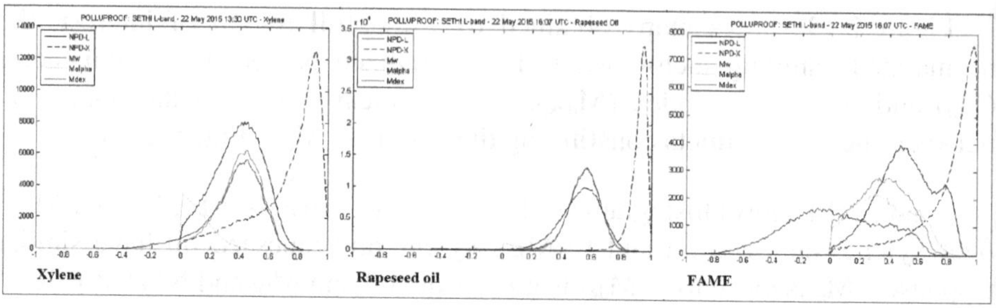

Figure 2.14. *Histograms: NPD L-band (black), NPD X-band (dash black), MW (red), Mα (green) and Mdex (blue) over xylene (left), rapeseed oil (center) and FAME (right). For a color version of the figure, see www.iste.co.uk/lefloch/remote.zip*

2.5. Conclusion

Due to the increase of maritime transport of hazardous and noxious substances (HNS), defined by the 2000 OPRC-HNS Protocol as substances other than hydrocarbon oil, which, if introduced into the marine environment, are likely to create hazards to human health, controlling chemical pollution at sea has become crucial. As for oil spills, remote sensing is of great interest for detecting, quantifying and characterizing chemical products that are discharged. However, our knowledge of the ability of remote sensing to achieve this is still limited.

An experimental campaign of acquisition (POLLUPROOF) was conducted in May 2015 over the Mediterranean Sea, during which controlled releases of hazardous and noxious substances were realized, the aim being to establish a procedure for collecting evidence of illegal maritime pollution by noxious and hazardous liquid substances using airborne sensors.

Among the six products tested during this experimentation at sea, three were detected without any ambiguity using SAR imagery: rapeseed oil, FAME and xylene. For the three others, the non-detectability can be caused either by a high volatility of tested products or by an impact on the water column that physically does not affect the backscattered signal at microwave frequency.

In addition, to demonstrate the ability of radar imagery to detect some HNS, two original results were presented in this chapter: first, an accurate method using X- and L-band radar imagery has been developed to detect and quantify the impact of chemical products at sea. X-band can easily detect the spills, even when the impact on the surface is limited; L-band is then used to quantify this impact.

A normalized polarization difference (NPD) parameter was then introduced for this purpose. We showed that, at L-band, the NPD parameter takes a wide range of values within the spill; this variation is related to the impact of the product on the ocean surface. Of course, this important result is only accessible to Earth observation systems whose instrumental noise level is low enough. We showed that the distinction between two HNS (with different thickness) within the same spill is possible with radar imagery only. We conclude that using SAR data with the suitable wavelength can allow us to define the characteristics of the slick along a spectrum ranging from thin surface films to thicker emulsions. This information is critical for efficient clean-up operations.

As future work, methods developed and tested on HNS will be tested on X- and L-band SAR data acquired by ONERA during the Norwegian oil spill experiment organized and managed by NOFO (Norwegian Clean Sea Association for Operating Companies) in June 2015.

2.6. Acknowledgments

The research presented in this chapter is part of the POLLUPROOF research program (ANR-13-ECOT-007) funded by the French National Research Agency (ANR). The authors are very grateful to everyone involved in the experiment at sea (ONERA, CEDRE, AVDEF, DGDDI and the French Navy) and those who participated in SAR data processing. They are also very grateful to G. Soriano and C.-A. Guerin for GO-SSA modeling and to P. Dubois-Fernandez and H. Oriot from ONERA for successful discussions.

2.7. References

[ANG 14] ANGELLIAUME S., FABBRO V., SORIANO G. *et al.*, "The GO-SSA extended model for all-incidence sea clutter modeling", *IGARSS Geoscience and Remote Sensing Symposium*, Quebec, Canada, July 13–18, 2014.

[ANG 15] ANGELLIAUME S., DUBOIS-FERNANDEZ P., MIEGEBIELLE V. *et al.*, "Polarimetric parameters for oil slicks detection using SAR data remote sensing – An evaluation", *IGARSS Geoscience and Remote Sensing Symposium*, Milan, Italy, July 26–31, 2015.

[BON 09] BONIN G., DUBOIS-FERNANDEZ P., RUAULT-DU-PLESSIS O. *et al.*, "The new ONERA multispectral airborne SAR system in 2009", *Radar Conference*, Pasadena, USA, May 4–8, 2009.

[BRE 05] BREKKE C., SOLBERG A.H.S., "Oil spill detection by satellite remote sensing", *Remote Sensing of Environment*, vol. 95, no. 1, pp. 1–13, 2005.

[ELF 97] ELFOUHAILY T., CHAPRON B., KATSAROS K. *et al.*, "A unified directional spectrum for long and short wind-driven waves", *Journal of Geophysical Research: Oceans*, vol. 102, no. C7, pp. 15781–15796, 1997.

[GAD 98] GADE M., ALPERS W., HÜHNERFUSS H. *et al.*, "Imaging of biogenic and anthropogenic ocean surface films by the multifrequency/multipolarization SIR-C/X-SAR", *Journal of Geophysical Research: Oceans*, vol. 103, no. C9, pp. 18851–18866, 1998.

[GAR 09] GARCIA-PINEDA O., ZIMMER B., HOWARD M. *et al.*, "Using SAR images to delineate ocean oil slicks with a texture-classifying neural network algorithm (TCNNA)", *Canadian Journal of Remote Sensing*, vol. 35, no. 5, pp. 411–421, 2009.

[GIR 05] GIRARD-ARDHUIN F., MERCIER G., COLLARD F. *et al.*, "Operational oil-slick characterization by SAR imagery and synergistic data", *IEEE Journal of Oceanic Engineering*, vol. 30, no. 3, pp. 487–495, 2005.

[GRE 12] GREGERS-HANSEN V., MITTAL R., "An improved empirical model for radar sea clutter reflectivity", *IEEE Transaction on Aerospace and Electronic Systems*, vol. 48, no. 4, pp. 3512–3524, 2012.

[GUE 10] GUÉRIN C.A., SORIANO G., CHAPRON B., "The weighted curvature approximation in scattering from sea surfaces", *Waves in Random and Complex Media*, vol. 20, no. 3, pp. 364–384, 2010.

[HAJ 03] HAJNSEK I., POTTIER E., CLOUDE S.R., "Inversion of surface parameters from polarimetric SAR", *IEEE Transactions on Geoscience and Remote Sensing*, vol. 41, no. 4, pp. 727–744, 2003.

[INT 11] INTERNATIONAL MARITIME ORGANIZATION, International Code for the Construction and Equipment of Ships carrying Dangerous Chemicals in Bulk (IBC Code), available at: http://www.imo.org/en/OurWork/Safety/Cargoes/CargoesInBulk/Pages/IBC-Code.aspx (accessed June 2020), 2011.

[KUD 12] KUDRYAVTSEV V.N., CHAPRON B., MYASOEDOV A.G. *et al.*, "On dual co-polarized SAR measurements of the ocean surface", *IEEE Geoscience and Remote Sensing Letters*, vol. 10, no. 4, pp. 761–765, 2012.

[LEI 12] LEIFER I., LEHR W., SIMECEK-BEATTY D. *et al.*, "State of the art satellite and airborne marine oil spill remote sensing: application to the BP Deepwater Horizon oil spill", *Remote Sensing of Environment*, vol. 124, pp. 185–209, 2012.

[MIN 12a] MINCHEW B., "Determining the mixing of oil and sea water using polarimetric synthetic aperture radar", *Geophysical Research Letters*, vol. 39, no. 16, L16607, 2012.

[MIN 12b] MINCHEW B., JONES C.E., HOLT B., "Polarimetric analysis of backscatter from the deepwater horizon oil spill using l-band synthetic aperture radar", *IEEE Transactions on Geoscience and Remote Sensing*, vol. 50, no. 10, pp. 3812–3830, 2012.

[SOL 12] SOLBERG A.H.S., "Remote sensing of ocean oil-spill pollution", *Proceedings of the IEEE*, vol. 100, no. 10, pp. 2931–2945, 2012.

[SOR 08] SORIANO G., GUÉRIN C.A., "A cutoff invariant two-scale model in electromagnetic scattering from sea surfaces", *IEEE Geoscience and Remote Sensing Letters*, vol. 5, no. 2, pp. 199–203, 2008.

[VAL 78] VALENZUELA G.R., "Theories for the interaction of electromagnetic and oceanic waves – A review", *Boundary-Layer Meteorology*, vol. 13, nos 1–4, pp. 61–85, 1978.

[WIS 98] WISMANN V., GADE M., ALPERS W. *et al.*, "Radar signatures of marine mineral oil spills measured by an airborne multi-frequency radar", *International Journal of Remote Sensing*, vol. 19, no. 18, pp. 3607–3623, 1998.

3

Remote Sensing of HNS using Longwave Infrared Hyperspectral Imaging

3.1. Introduction

A longwave infrared (LWIR) hyperspectral imager was deployed in support of the POLLUPROOF project for the detection and identification of HNS in a maritime environment. In October 2014, the Hyper-Cam sensor was installed on an articulated platform at a height of 12 m above a seawater basin at CEDRE where several HNS were spilled. Several gases/liquids, including methanol, diethyl ether, xylene and toluene, were detected and identified from the exploitation of longwave infrared imagery using the Generalized Likelihood Ratio Test (GLRT). In May 2015, the Hyper-Cam was installed onboard a Cessna C-303 and flown at an altitude of 7,500 ft above sea level in order to detect and identify spills of HNS in the Mediterranean Sea. The sensor was capable of detecting several slicks, such as colza oil and fatty acid methyl ester, but to date, it has not been possible to identify the chemical material forming the slick. It should be noted that hyperspectral remote imagers operating in other bands (e.g. reflective band at 0.4–2.5 micron) may be better suited for some HNS spills.

With ever-increasing amounts of marine traffic involving vessels carrying larger quantities of hazardous and noxious substances (HNS) worldwide, it is important that techniques be developed to detect these chemicals in the event of maritime spills. Some of today's largest ocean-going chemical tankers can

Chapter written by Eldon PUCKRIN, Dennis DURO, Guillaume GAGNE, Anne-Pier BERNIER, Louis ARMSTRONG and Sophie CHATAING.

carry as much as 100,000 DWT in cargo, and any major mishap could result in an environmental disaster along coastal areas or at major ports. Several methods already have been developed for the detection of spills involving petroleum oil, which lies outside the definition of an HNS. These methods include Side-Looking Airborne Radar (SLAR) [WAG 00] and remote sensing methods based on IR and UV line scanner techniques [FIN 11]. It is equally important that effective detection methods be developed for HNS in a maritime environment.

In this chapter, the deployment of a longwave infrared (LWIR) hyperspectral remote sensing imager during the POLLUPROOF project in 2014 and 2015 was examined to determine whether it is feasible to detect some of the more important HNS threats in a maritime environment. The focus of this technique on detecting the unique spectral signature of a chemical may make it a useful technology for detecting and identifying a wide assortment of HNS. In the POLLUPROOF project, a small selection of HNS were chosen for investigation, including methanol, toluene and xylene, which have infrared signatures in the 8–12 micron longwave infrared band. Results of the detection and identification of these HNS are presented in the following sections.

3.2. LWIR hyperspectral remote sensing capability

Most commercial off-the-shelf LWIR hyperspectral imaging sensors are based on the dispersive push-broom-type or an interferometer-based configuration. In this project, the Hyper-Cam sensor (Telops, Inc.), which is based on the latter configuration, was used to detect HNS spills in a maritime environment. The sensor, shown in Figure 3.1, operates in the 8–12 micron thermal infrared region and incorporates a 320×256 pixel focal plane array detector based on cooled mercury–cadmium–telluride. In this work, the sensor was configured with a 0.35 mrad spatial resolution and at a spectral resolution of 8 cm^{-1}. The instrument has two integrated calibration blackbodies used to perform a radiometric calibration of the infrared measurements. The sensor also has advanced acquisition and processing electronics that offer the capability of converting the raw interferograms into spectra using a real-time discrete Fourier transform. The sensor can be either deployed in a ground-based configuration or mounted on a stabilization platform for integration in an aircraft. The Hyper-Cam has been used in several ground-based and airborne measurement campaigns previously [FAR 07, PUC 10, TUR 13].

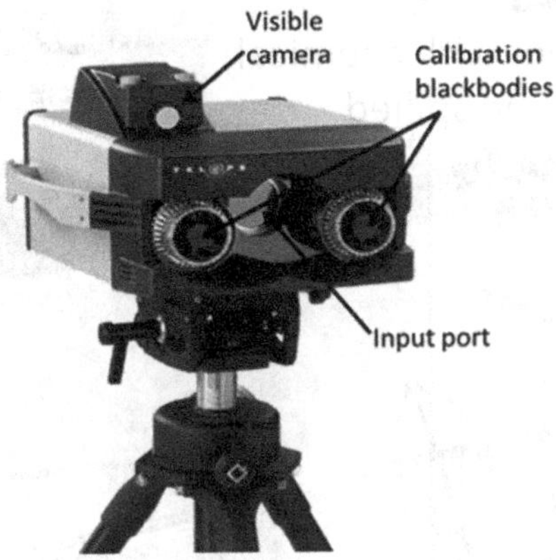

Figure 3.1. *Photo of the Hyper-Cam LWIR imaging sensor*

3.2.1. *Basin measurements at CEDRE*

During HNS spill experiments at the CEDRE basin in October 2014, the Hyper-Cam was installed on a boom lift platform at a height of 12 m above ground level, as shown in Figure 3.2.

A sequence of hyperspectral measurements were made of the spill area, shown in Figure 3.3, that started before the spill began and continued for a period of several minutes after the spill occurred to determine the behavior of the resulting gas plume/slick. Details of the spill releases at the CEDRE basin can be found in a companion reference [CHA 16].

Figure 3.2. *(a) Hyper-Cam LWIR imaging sensor installed on platform of a boom lift looking straight down. A HySpex hyperspectral imager covering the 0.4–2.5 micron region was also installed and operated by personnel from ONERA. (b) Hyper-Cam installed on boom lift 12 m above ground*

Figure 3.3. *Region of the spill area viewed by the Hyper-Cam, and the container with the chemical to be spilled into the seawater by CEDRE personnel*

3.2.2. *Sea measurements*

The Hyper-Cam has been successfully used previously [PUC 10] in an airborne configuration onboard a fixed-wing aircraft using a stabilization platform. The complete system is shown in Figure 3.4.

Figure 3.4. *Airbome platform for flying the Hyper-Cam instrument in a fixed-wing aircraft [PUC 10]*

The airborne configuration includes a stabilization platform, the Hyper-Cam instrument, an Image Motion Compensation (IMC) mirror, an INS/GPS unit, a laser range finder and one visible boresight camera. All these modules are rigidly mounted on a high-stiffness base plate. The IMC mirror is used to compensate the airplane pitch, roll and forward motion, while the stabilization platform is used to reduce the airplane vibrations and to compensate the airplane yaw. The IMC mirror is controlled by the navigation module, which receives and uses the information from a laser range finder, for altitude determination, and a GPS/INS unit. The GPS/INS also enables ortho-rectification and geo-referencing of the collected data. In the airborne configuration, the Hyper-Cam performs a step-and-stare measurement, as shown in Figure 3.5, where the pointing mechanism sweeps in order to provide a constant ground field of view (FOV), i.e. a staring mode. The pointing mechanism then returns to its original position in order to be ready to observe the next ground FOV. A percentage of the FOV is used for overlap between two consecutive ground FOVs to ensure no ground area is missed. An interferometer-based imager, such as the Hyper-Cam, requires the measured scene to be stable during the full measurement of the interferogram, which usually requires about 1 s, depending on the size of the region being imaged.

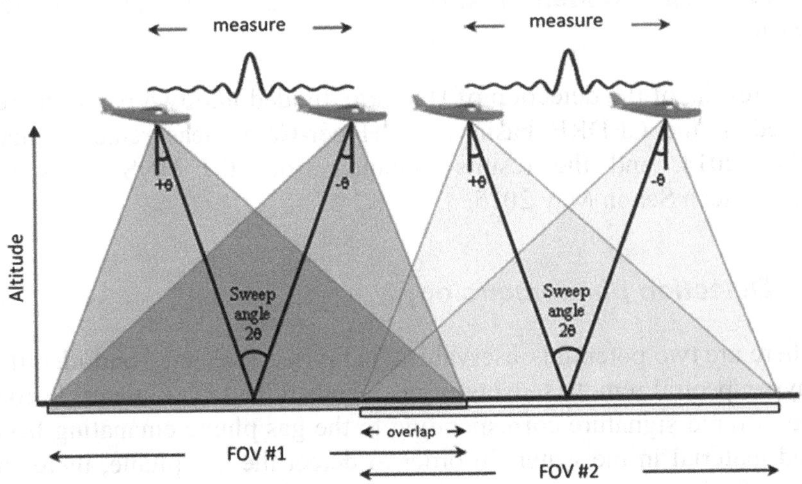

Figure 3.5. *Step-and-stare airborne measurement configuration for an interferometer-based imager [PUC 10]*

Airborne measurements of chemical spills carried out in May 2015 in the Mediterranean Sea off the coast of Toulon were performed with this system configuration installed onboard a Cessna C-303 aircraft, as shown in Figure 3.6. Details of the spill releases at sea can be found in another reference [CHA 16].

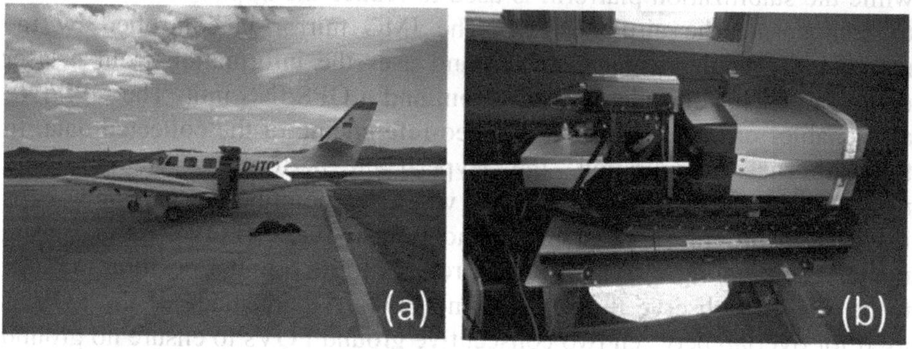

Figure 3.6. *(a) Cessna C-303 aircraft deployed with (b) Hyper-Cam airborne system installed inside looking down through an open hole*

3.3. Detection and identification of HNS using LWIR hyperspectral sensing

The results of the detection of HNS are divided into two parts: the results obtained at the CEDRE basin from Hyper-Cam measurements made in October 2014, and the results obtained from the HNS spills in the Mediterranean Sea in May 2015.

3.3.1. *Detection phenomenology*

There are two potential observables that may be detected and identified in the hyperspectral remote sensing of chemical spills. One observable consists of the infrared signature corresponding to the gas plume emanating from the spilled material in the water. In order to detect the gas plume, there should be a sufficient concentration of the gas and a sufficient temperature difference between the plume and the background scene.

An expression [LAV 05] that approximates the difference in the radiance (δL) associated with a gas plume (having transmittance τ_{gas}) and a clear scene without gas present (L_{clear}) is given by:

$$\delta L = (1 - \tau_{gas})(B(T_{gas}) - L_{clear}) \tag{3.1}$$

where $B(T_{gas})$ is the Planck radiance at the temperature of the gas. The first term in the expression shows the dependence of the radiance on the transmittance of the gas, and the second term shows the dependence on the difference in temperature between the gas and background. Figure 3.7 shows the absorbance spectra of some of the gas signatures used in the experiments at CEDRE. These include signatures for methanol, xylene and toluene; however, other liquids including diethyl ether were also spilled to test the detection procedure. The absorbance intensity of the signatures varies by an order of magnitude between toluene and methanol, and with different vapor pressures, these chemicals provide interesting test cases for the detection capability.

The second possible observable, which consists of a thin liquid film on the surface of the water, requires the use of the Fresnel equations to describe the detection phenomenon [BOR 99].

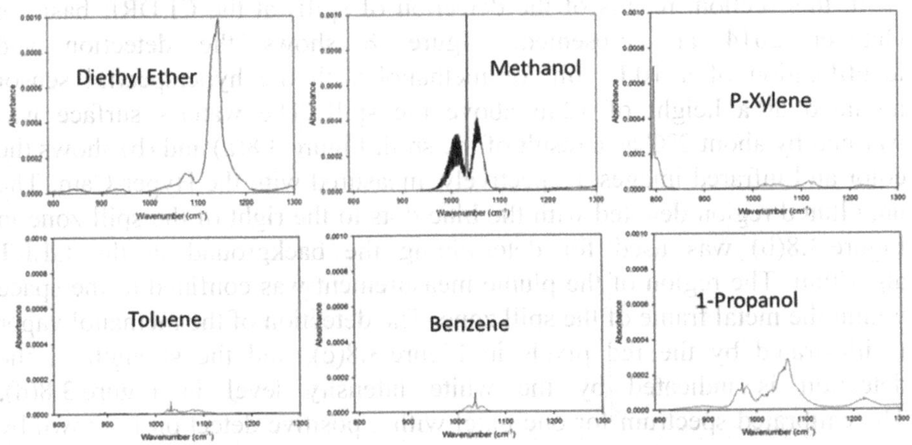

Figure 3.7. *Absorbance signatures of various chemicals spilled at the CEDRE basin*

3.3.2. *Detection algorithm*

For both the basin and sea measurements, the hyperspectral images were exploited using the Generalized Likelihood Ratio Test (GLRT). A detailed description of the detection method is given in a previous work [MAN 03]. The GLRT detection algorithm was used to identify the gas plumes (or slicks) originating from the chemical spills in the water. Essentially, the GLRT is described mathematically by:

$$GLRT \quad \frac{m^T P_B^{\perp} m}{m^T P_{BS}^{\perp} m} = \frac{(P_B^{\perp} m)^T (P_B^{\perp} m)}{(P_{BS}^{\perp} m)^T (P_{BS}^{\perp} m)} = \frac{\|P_B^{\perp} m\|^2}{\|P_{BS}^{\perp} m\|^2} \qquad [3.2]$$

where the numerator represents the squared norm of the measurement, *m*, projected out of the background space, *B*, and the denominator represents the squared norm of the measurement projected out of the background + signature space, *S*. If the measurement does not contain the signature of interest, the result of the GLRT is approximately 1. If the measurement does contain the signature of interest, the GLRT will produce a value greater than 1.

3.3.3. *Basin measurements at CEDRE*

In this section, results of the detection of spills at the CEDRE basin in October 2014 are presented. Figure 3.8 shows the detection and identification of a 10 L spill of methanol with the hyperspectral sensor mounted at a height of 12 m above the spill. The water's surface was warmed by about 2°C as a result of the spill. Figure 3.8(a) and (b) shows the color and infrared images, respectively, measured with the Hyper-Cam. The unpolluted region denoted with the blue dots to the right of the spill zone in Figure 3.8(b) was used for determining the background in the GLRT algorithm. The region of the plume measurement was confined to the space within the metal frame of the spill zone. The detection of the methanol vapor is illustrated by the red pixels in Figure 3.8(c), and the strength of the detection is indicated by the white intensity level in Figure 3.8(d). The calibrated spectrum for one pixel with a positive detection is shown by the blue spectrum in Figure 3.8(e). This is compared to the reference spectrum of pure methanol vapor (from Figure 3.7). It is clear that even in the unfiltered measurement, the presence of methanol is evident from the spectral features. Figure 3.8(f) shows the filtered spectral output of the GLRT detector. Essentially the projected measurement result agrees with the projected methanol signature (taking into account a phase shift of 180 degrees in the display).

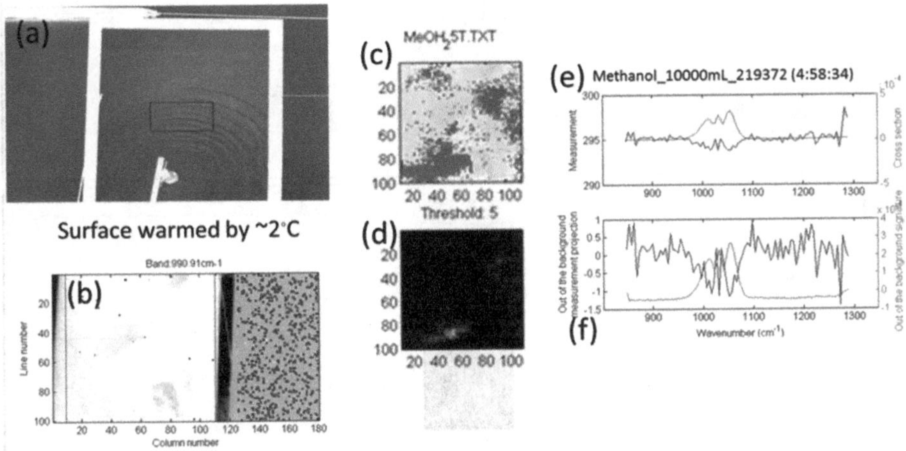

Figure 3.8. *(a) Visible image of the spill confinement zone; (b) Hyper-Cam infrared image of confinement zone immediately after spill; (c) red pixels indicate spill vapor detections within the spill zone; (d) the detection plane shows the strength of the detection (stronger detections are shown by brighter pixels); (e) the Hyper-Cam measurement (blue) for one pixel in the spill zone and the methanol reference signature (red); and (f) the projected results from the GLRT detector. For a color version of the figure, see www.iste.co.uk/lefloch/remote.zip*

Figure 3.9 shows the detection of a 5 L spill of diethyl ether for two different times. In this case, the surface of the water was cooled by about 2°C. Figure 3.9(a) displays a similar series of graphs as described for Figure 3.8. The diethyl ether is clearly still present, but starting to subside from the spill zone. Again, it is clear that even in the unfiltered spectral measurement in the right side of the panels, the presence of diethyl ether is evident from the spectral features. The filtered spectral output of the GLRT detector also shows very good agreement between the projected measurement result and the diethyl ether signature.

Figure 3.10 shows the detection of a 5 L spill of p-xylene. Using a similar analysis as discussed for Figure 3.8, it appears that the vapor from the p-xylene spill is also detected, as shown in Figure 3.10(c), although it is clear that the level of agreement in the filtered results of Figure 3.10(f) is less satisfactory than for the other spills.

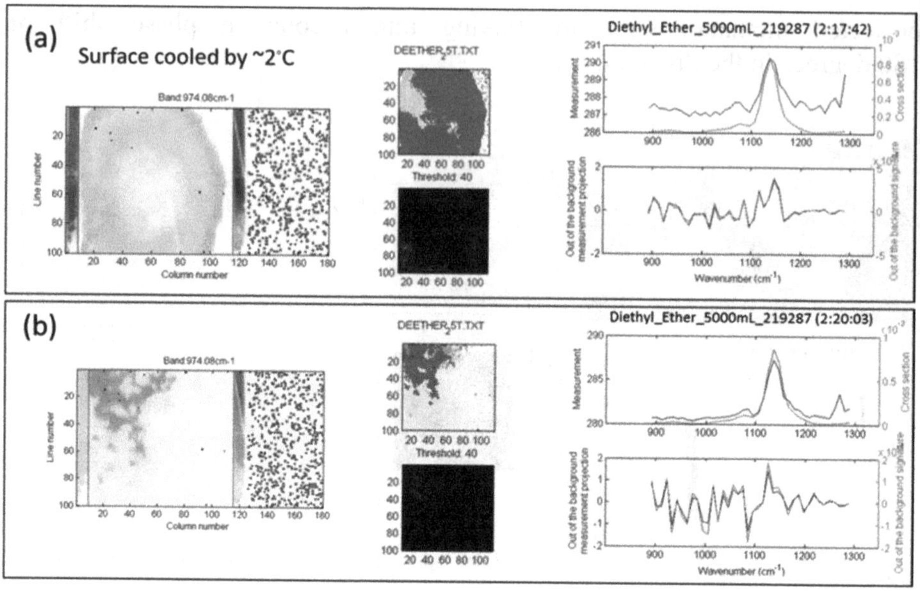

Figure 3.9. *(a) Detection results from a 5 L spill of diethyl ether, similar to that discussed for methanol in Figure 3.8; (b) a subsequent detection result after a time delay of 2.5 min. For a color version of the figure, see www.iste.co.uk/lefloch/remote.zip*

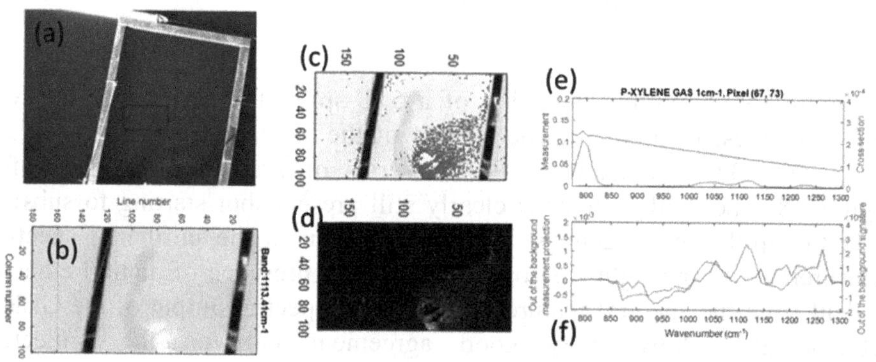

Figure 3.10. *(a) Visible image of the spill confinement zone, for p-xylene; (b) Hyper-Cam infrared image of confinement zone immediately after spill; (c) red pixels indicate p-xylene vapor detections within the spill zone; (d) the detection plane shows the strength of the detection (stronger detections are shown by brighter pixels); (e) the Hyper-Cam measurement (blue) for one pixel in the spill zone and the p-xylene reference signature (red); and (f) the projected results from the GLRT detector. For a color version of the figure, see www.iste.co.uk/lefloch/remote.zip*

Figure 3.11 shows the detection of a 5 L spill of toluene. In Figure 3.11(a), the mechanism to release the spill is visible in the image. Using a similar analysis as discussed for Figure 3.8, it appears that the vapor from the toluene spill is detected, as shown in Figure 3.11(c).

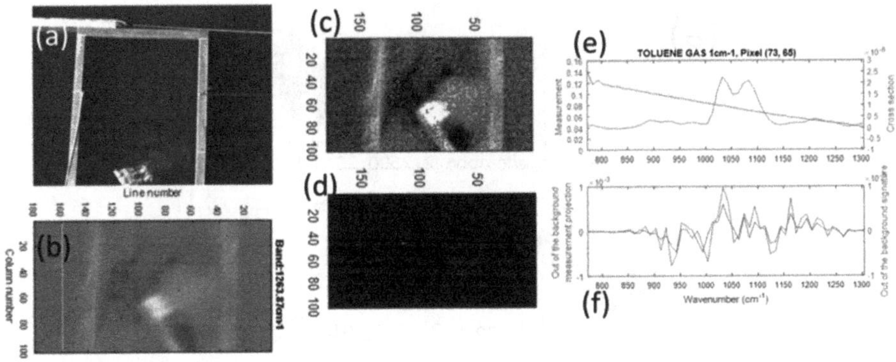

Figure 3.11. *(a) Visible image of the spill confinement zone for toluene; (b) Hyper-Cam infrared image of confinement zone immediately after spill; (c) red pixels indicate toluene vapor detections within the spill zone; (d) the detection plane shows the strength of the detection (stronger detections are shown by brighter pixels); (e) the Hyper-Cam measurement (blue) for one pixel in the spill zone and the toluene reference signature (red); and (f) the projected results from the GLRT detector. For a color version of the figure, see www.iste.co.uk/lefloch/remote.zip*

For all of the detection results in Figures 3.8–3.11, it is possible that the signature of the liquid spill is also contributing to a positive detection result. This is especially possible for spills involving xylene that exhibit low vapor pressures and where the liquid signature may be the only observable present to be detected. In general, the infrared signatures that correspond to liquid features are similar to their gaseous counterparts, although they may be shifted spectrally by a few wavenumbers and may exhibit a somewhat different lineshape. The refraction by the thin liquid film may also affect the detection, but generally for films of thickness less than 2 microns, these effects are not large. Additional work is being undertaken to fully understand the contribution of vapor and liquid signatures on the detection results of these spills.

3.3.4. *Sea measurements*

Hyper-Cam images for the third release consisting of colza oil and fatty acid methyl ester (FAME) that took place on May 22, 2015, from 15:20 to

16:50 UTC, were exploited with the GLRT detector. Two infrared images for colza oil and FAME are shown in Figure 3.12. The images were collected with the sensor installed in a Cessna C-303 flying at an altitude of 7,500 ft above the water.

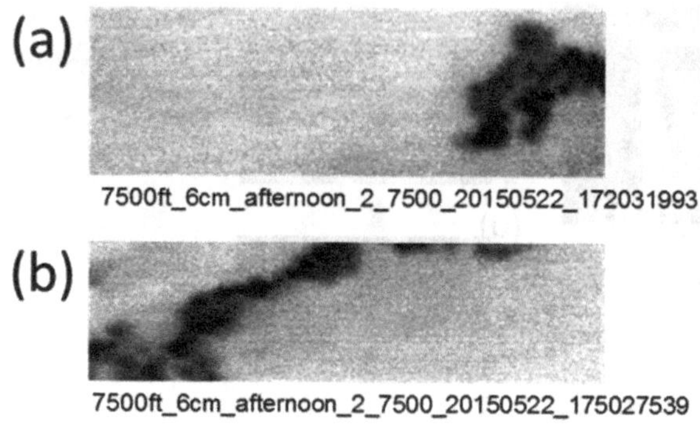

Figure 3.12. *(a) Infrared image of a colza oil slick measured at 15:20 UTC (17:20 LT); (b) a FAME slick measured at 15:50 UTC (17:50 LT)*

The results of the GLRT detection are shown in Figure 3.13. In this case, there is confusion between the identities of the slicks. Either the colza or the FAME is detected for both slicks, which is probably due to the weak signatures of these materials in the LWIR region. Work on detecting the xylene slick that was released earlier in the afternoon is currently being investigated.

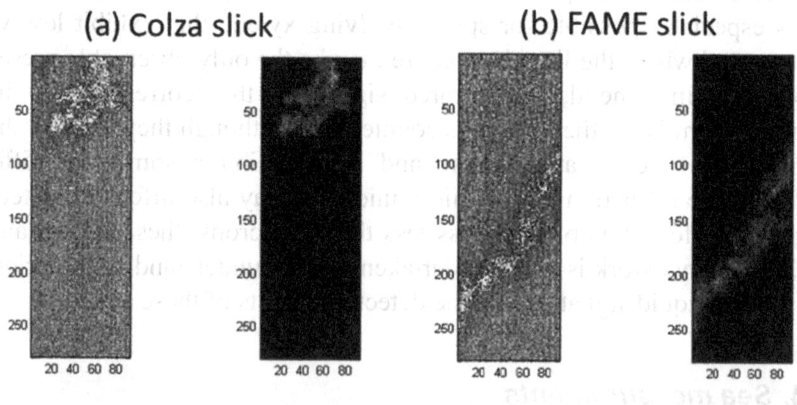

Figure 3.13. *GLRT detection results of (a) colza oil slick and (b) FAME slick. The slicks are detected but not successfully identified in the LWIR band. For a color version of the figure, see www.iste.co.uk/lefloch/remote.zip*

3.4. Conclusion

Longwave infrared hyperspectral remote sensing of HNS using the Hyper-Cam sensor in both ground-based and airborne configurations was undertaken during the POLLUPROOF project during October 2014 (CEDRE basin) and May 2015 (Mediterranean Sea). The hyperspectral images of the spills of several different materials at CEDRE were analyzed with the GLRT detector. Several gases/liquids were detected, including methanol, diethyl ether, xylene and toluene. Other materials such as colza were much more difficult to detect and identify in the longwave infrared band due to the lack of a strong spectral signature. It should be noted that hyperspectral remote sensing in other bands (e.g. reflective 0.4–2.5 micron) may be better suited for some HNS spills.

Exploitation of colza oil and FAME spills in the Mediterranean Sea showed that the slicks may be detected, but not yet identified with the Hyper-Cam sensor. Additional work is being performed to understand the spectral signatures of these chemicals and the exploitation methodology.

3.5. References

[BOR 99] BORN M., WOLF E., *Principles of Optics: Electromagnetic Theory of Propagation, Interference and Diffraction of Light Hardcover*, Cambridge University Press, 1999.

[CHA 16] CHATAING S., ANGELLIAUME S., FOUCHER P.Y. *et al.*, "POLLUPROOF Project: PROOF improvement of HNS maritime pollution by airborne radar and optical facilities – Experimental approach", *Sea Tech Week – Sea and Digital Technologies*, Brest, France, October 10–14, 2016.

[FAR 07] FARLEY V., CHAMBERLAND M., LAGUEUX P. *et al.*, "Chemical agent detection and identification with a hyperspectral imaging infrared sensor", *Proceedings of SPIE*, vol. 6661, p. 66610L, 2007.

[FIN 11] FINGAS M. (ed.), *Oil Spill Science and Technology: Prevention, Response and Clean-up*, Elsevier, Gulf Professional Publishing, Boston, 2011.

[LAV 05] LAVOIE H., PUCKRIN E., THERIAULT J.M. *et al.*, "Passive standoff detection of SF6 at a distance of 5.7 km by differential fourier transform infrared radiometry", *Applied Spectroscopy*, vol. 59, no. 10, pp. 1189–1193, 2005.

[MAN 03] MANOLAKIS M., MARDEN D., SHAW G.A., "Hyperspectral image processing for automatic target detection applications", *Lincoln Laboratory Journal*, vol. 14, no. 1, pp. 79–116, 2003.

[PUC 10] PUCKRIN E., TURCOTTE C.S., LAHAIE P. *et al.*, "Airborne infrared-hyperspectral mapping for detection of gaseous and solid targets", *Chemical, Biological, Radiological, Nuclear, and Explosives (CBRNE) Sensing XI,* May 5, 2010.

[TUR 13] TURCOTTE C.S., PUCKRIN E., AUBÉ F. *et al.*, "Novel compact airborne platform for remote sensing applications using the Hyper-Cam infrared", *Airborne Intelligence, Surveillance, Reconnaissance (ISR) Systems and Applications X,* Baltimore, United States, 2013.

[WAG 00] WAGNER P., HENGSTERMANN T., ZIELINSKI O., "MEDUSA: an airborne multispectral oil spill detection and characterization system", *Infrared Technology and Applications XXVI,* 2000.

PART 2

Remote Sensing to Support Marine Surveillance Services

4

Customs Expertise in Remote Sensing

4.1. Introduction

In order to contribute to the mission of surveillance and law enforcement in marine pollution, the French Customs have two dedicated aircraft for remote sensing of marine pollution (F406 POLMAR III). These air assets are in charge of collecting surface information with their airborne remote sensing tools.

Primarily, these specific remote sensing equipment were developed to observe oil release.

The crew members were trained, including by CEDRE, to detect, identify and characterize marine pollution. This observation alone is enough to recognize pollution and determine the legality or illegality of a leakage.

4.2. The aircraft

French Customs have two dedicated aircraft for remote sensing of marine pollution:

– A Reims-Aviation F406 POLMAR III, fitted with a side-looking airborne radar (SLAR), a scanner (infrared/UV and visible optical) and an MX10 camera. The F-ZBGD is based at Lorient Lann-Bihoué Air Base.

Chapter written by Laurent BUIGNET.

Figure 4.1. *The F406 POLMAR III*

– A Reims-Aviation F406 POLMAR II (retrofitted in POLMAR III), fitted with a side-looking airborne radar (SLAR), a scanner (infrared/UV and visible optical) and a microwave radar. The F-ZBCG is based at Hyères Air Base. This aircraft is fitted with a specific microwave radar that is able to determine the thickness of the wake pollution (by comparing the difference in temperature with the seawater).

Figure 4.2. *The F406 POLMAR II*

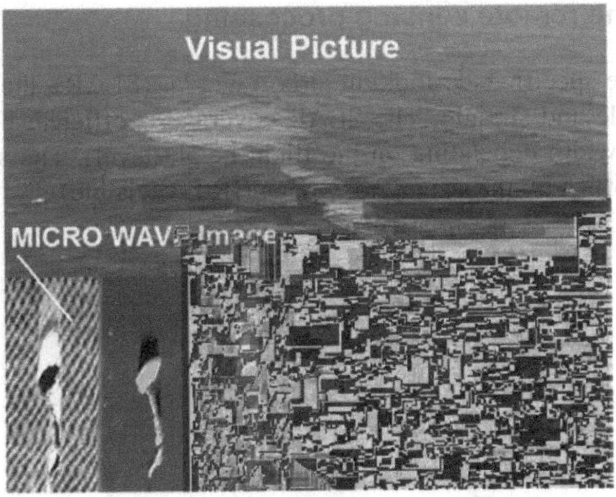

Figure 4.3. *Comparison between the visual picture and microwave images. For a color version of the figure, see www.iste.co.uk/lefloch/remote.zip*

4.3. The equipment

Inside the two F406 POLMAR III, we find only one computer with a 17-inch screen which brings together the different functions: remote sensing images, map, photos and video processing, and draft reports.

Figure 4.4. *Computer inside the F406 POLMAR III*

4.4. Airborne remote sensing processing

Due to the specific side-looking airborne radar (SLAR), the flight profile must be stable and straight. The SLAR coverage is efficient between 2 Nm and 20 Nm (width depending on the flight height/level). The sensor hole in the plane is offset by the scanner coverage (IR/UV/visible).

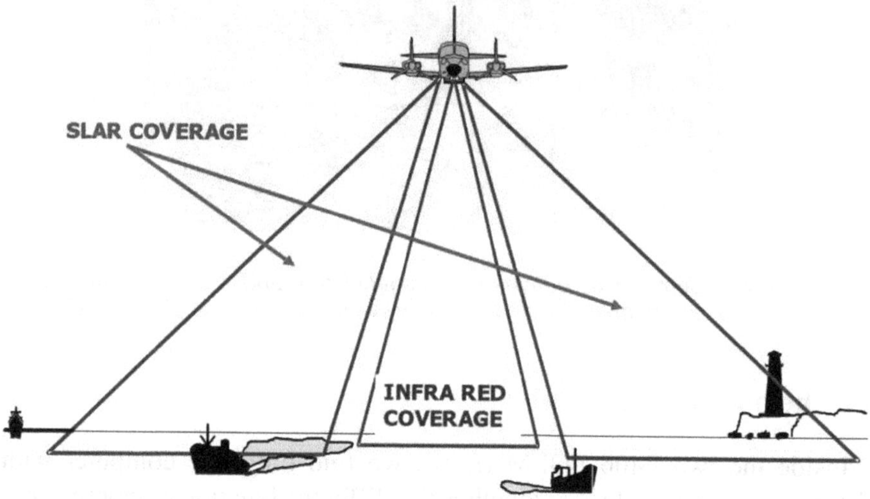

Figure 4.5. *SLAR and IR coverage. For a color version of the figure, see www.iste.co.uk/lefloch/remote.zip*

4.5. Side-looking airborne radar (SLAR) processing

This radar detects abnormalities in the sea surface, due to an oil or HNS product that removes the wave capillarity at the water surface.

This system has some limitations for use at sea, which are: 0/1<SEA CONDITIONS<4/5.

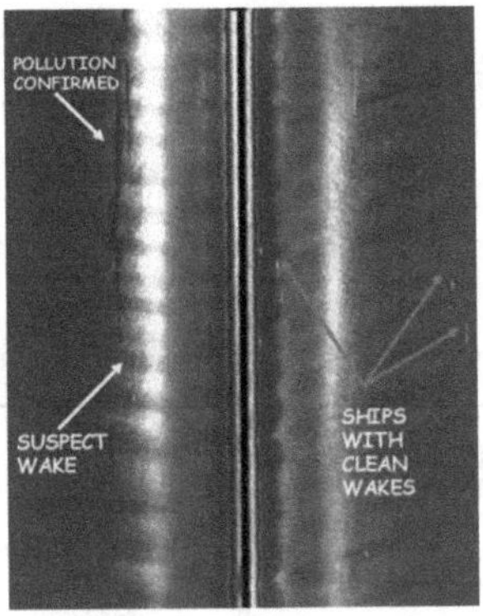

Figure 4.6. *SLAR image*

4.6. Infrared and ultraviolet line scanner

The IR detects infrared radiation with a wavelength in the band of 8–12 μm emitted from the oil. These oil layers radiate more slowly than the surrounding clear sea, and show up as variations in gray levels (or in defined colors). Thicker layers (greater than about 0.5 mm) will absorb sunlight more rapidly than the surrounding sea and show white on the display.

An ultraviolet scanner is a passive device that detects reflected ultraviolet with a wavelength of about 0.3 μm.

4.7. Standard detection and investigation

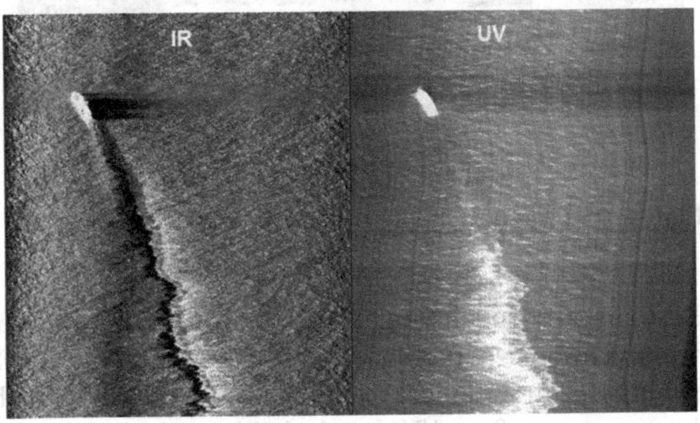

Figure 4.7. *IR and UV images*

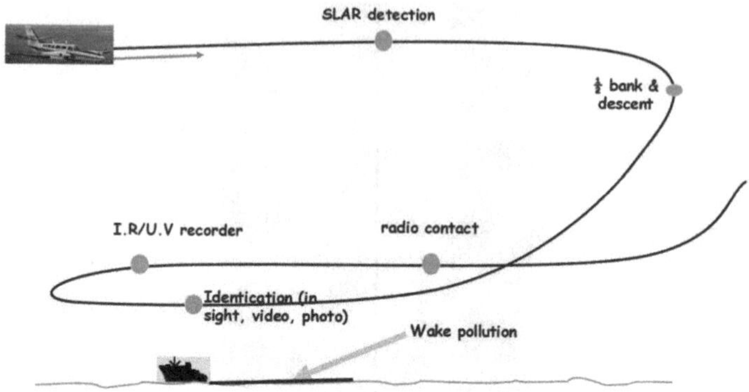

Figure 4.8. *Standard detection process*

In flight, our mission is to closely monitor, detect, investigate, identify and report any illegal discharges from suspected polluters, in compliance with national and MARPOL regulations, in order to collect evidence for the prosecution of offenders.

4.8. The future, a new multi-mission aircraft

Figure 4.9. *The new maritime patrol aircraft in development*

French Customs are developing a new maritime patrol KA350ER-FC aircraft with conventional surveillance systems (panoramic OM400 Radar and infrared multi-spectral camera) and with a side-looking airborne radar (SLAR) and a scanner (infrared/UV and visible optical) for remote sensing of marine pollution.

5

Remote Sensing as Evidence in Court

5.1. Introduction

This chapter is based on an oral presentation made by the author at the CEDRE Information Day on October 16, 2016, and should therefore not be considered as an article on legal doctrine, which would call for more substantial bibliographical references together with references to past cases. The conclusions of this chapter are based solely on the experience and analysis of the author, an operational legal expert and research associate. The complex question of the future use of remote sensing equipment to provide evidence in court requires, in our opinion, a comprehensive, in-depth study in comparative, international and European law in order to obtain actionable conclusions.

This chapter is divided into four sections: first, an introduction to the issue of "remote sensing and evidence", then a section presenting the legal frameworks of offences and of evidence in terms of operational discharge at sea (section 5.2), followed by a section on remote sensing, how it is considered by the courts and questions and advances observed in this field (section 5.3), and finally some conclusions and recommendations. This chapter focuses solely on the question of detecting and gathering evidence of illegal, deliberate discharge of polluting substances from ships, with regard to French law.

From a legal point of view, the question of remote sensing and its use to provide evidence calls for the definition and clarification of terms and notions relating to the technology, how it is implemented and its purpose.

Chapter written by Yann RABUTEAU.

The first point of clarification required is no doubt to determine what the purpose of remote sensing is, in general terms. In this respect, we understand that remote sensing is a means of detecting, measuring and acquiring information remotely, using instruments, systems, whether active or passive, without systematic or direct human involvement in implementing the equipment or observing the target. This initial point is of particular importance as it draws attention to the predominant role of the technological equipment in relation to direct observation "by the human eye".

Now, the equipment itself must be addressed [PIN 17]. The wide range of remote sensing technologies and measurement instruments (applicable to marine pollution) is striking: whether it be SLAR, UV scanners, FLIR cameras, satellite imagery or even digital videos and photographs, the variety is such that all the expected information can be made available, yet expertise is essential to interpret the information gathered in terms of evidence.

In this respect, what information is liable to be provided by remote sensing? We know that the existing solutions can be used to both detect and characterize discharges at sea. This is a fundamental point as the information on the type of spill observed, whether it is oil or another substance, an estimation of the surface area and thickness of the slick and the concentration of pollutant in the discharge are expected elements in terms of evidence, but they are also called for by the trial court and the parties to the criminal proceedings to gauge its severity.

Finally, it is important to clarify the notion of "evidence" itself in the field of deliberate discharge at sea. In French law, illegal discharge of oil or other polluting substances at sea from a ship (see section 5.2) constitutes a punishable offence. It is therefore governed by criminal law (definition, sanctions, jurisdictional rule), meaning that evidence of both the material element and the intentional element of the offence must be provided to condemn the offender. In terms of illegal discharge, these two points are of significant importance. Providing evidence of the material element implies here demonstrating the reality of the discharge in hand, which is in violation of the rules of the MARPOL Convention: for instance, demonstrating that the concentration of a reported discharge is indeed in excess of the 15 ppm concentration limit, in the case of a substance covered by Annex I of the convention. The intentional element, on the other hand, means that the deliberate nature of the discharge observed must be proven in order to distinguish it from an incident or accident.

If these two elements are demonstrated, the offence is constituted and criminal liability may be sought.

5.2. Legal framework of the offence and the evidence

5.2.1. *What the texts say*

We know that discharging polluting substances from a ship into the sea in violation of the rules of the MARPOL Convention constitutes an offence according to French law: a deliberate marine pollution offence. Today, all the annexes of the MARPOL Convention have been transposed into the French Environmental Code, as a basis for incrimination, hence:

– Article L.218-11: "Committing a discharge of a polluting substance in violation of the provisions of rules 15 and 34 of Annex I, relating to the prevention of pollution by oil, or an infringement of the provisions of rule 13 of Annex II, relating to the control of pollution by noxious liquid substances in bulk, of the MARPOL Convention is punishable, for any captain of a French ship, by a fine of 50,000 Euros."

– Article L.218-14: "Releasing harmful substances transported in packages in violation of the provisions of rule 7 of Annex III of the MARPOL Convention is punishable, for any captain of a French ship, by seven years of imprisonment and a fine of 1 million Euros."

– Article L.218-15:

- "I. Committing one of the offences described in the provisions of rule 11 of Annex IV and of rules 3, 4 and 5 of Annex V of the MARPOL Convention is punishable, for any captain of a French ship, by one year of imprisonment and a fine of 200,000 Euros."

- "II. Committing one of the offences described in the provisions of rules 12, 13, 16 and 18 of Annex VI of the MARPOL Convention, as well as in the provisions of I to IV of article L. 218-2, is punishable, for any captain of a French ship, by one year of imprisonment and a fine of 200,000 Euros."

It is also interesting to note that the level of monetary sanctions has evolved to reach a maximum fine of €15 million, since the law on environmental liability in 2008, for a discharge in violation of Annex I of the MARPOL Convention (see Article L.218-12 of the French Environmental Code).

Thus, the stakes are high and the question of evidence is taking on growing importance, in particular in light of the diversion, detainment and bond payment procedures, which are implemented following the observation of a discharge and the drafting of the official report by the agent.

The definition of the offence is therefore part of a global procedure and must be in compliance with international law, but no specific information is provided as concerns the evidence accepted for such offences. What do the texts say in this respect? To determine the legal specifications concerning evidence, we turn to general criminal law.

In criminal law, the choice of evidence is said to be "free". Under Article 427 of the French Code of Criminal Procedure: "Except where the law provides otherwise, offences may be established by any type of evidence and the judge shall decide based on his personal conviction. The judge shall base his decision solely on the evidence presented to him during the debates and discussed by the parties before him".

In other words, in the absence of express provisions on the type of evidence required and how it is adduced – which is the case of illegal discharge – evidence can be provided by all appropriate means. This provision would *a priori* appear to open the way to remote sensing but does not constitute the sole legal framework to be taken into consideration. Indeed, over and above this general provision as concerns evidence, an important point of clarification elucidated by the criminal system relating to ship-source discharge. This point relates to the value of the official report drawn up by the agent who observed the offence. The French Environmental Code states (in Article L.218-28) that: "The official reports drawn up by the agents mentioned in article L. 218-26 shall be valid until proven otherwise. They shall be immediately submitted to the Chief Prosecutor by the agent who reported the violation (…)".

The report has probative value and is therefore recognized and protected by law. The defendant must therefore provide evidence that the observations mentioned in the report are inaccurate.

Nevertheless, these two points (freedom of evidence and value of the report) are the only elements specified by the texts.

5.2.2. *What legal precedents have been set?*

In the absence of other legal provisions, many useful clarifications have been provided in terms of evidence of the offence by legal precedents in criminal cases of illegal discharge at sea.

In this respect, reference should be made to the *Traquair* case and the decision of the Rennes Court of Appeal on March 18, 1996 [ROY 06]. Without going into the details of this case of operational discharge, it is important to remember that 20 years ago (1996) court practice in this field was no doubt not as advanced as it is today, and the question of evidence of the material element of the illegal nature of the discharge remained to be clarified. In this respect, the decision of the Rennes Court of Appeal is noteworthy in that it makes explicit reference to the Bonn Agreement and its Oil Appearance Code. The Court stated that:

> "Given that a sample is not required by law as evidence of the offence, which can also be provided by direct visual observation from aircraft (...) corroborated by colour photographs (...) however, given that various experimentations (...) have demonstrated:
>
> – That no discharge with an oil content less than or equal to 15 mg/l has ever been detected visually,
>
> – That discharges with an oil content (...) of between 15 and 100 mg/l cannot be observed from an aircraft (...) given that they never form continuous films of over a few decimeters long".

It can hence be concluded that if the discharge can be detected visually, it is necessarily a violation of Annex I of the MARPOL Convention, as it exceeds the maximum authorized concentration. Since this legal precedent was set, further cases have followed and the elements expected to support observations have been clarified: photographs to support references to the Bonn Agreement, showing the sea conditions, the source of the discharge, the identity of the ship, etc. All these elements provided by photographs corroborate the direct visual observation by the agent in charge of drawing up the report.

According to case law, the intentional element, i.e. the deliberate nature of the observed discharge, is established based on the elements provided by the enquiry, the inspection of the ship, the hearing of the defendant and the different witnesses, as well as a "body of concurring evidence" (French Supreme Court, *Fast Rex* case, 18/03/2014).

5.3. Remote sensing: questions and advances

5.3.1. *Does the verdict of the* Traquair *case exclude recourse to remote sensing?*

This is the first question to be addressed from a legal point of view as, while this decision legitimized the use of the Bonn Agreement Oil Appearance Code, it also underlined the importance of direct visual observation. So what happens if the discharge cannot be directly and visually observed to be compared with this appearance code? If the discharge is only determined to be in excess of the concentration limit by remote sensing? If the discharge is observed at night? All these questions appear to be at odds with the requirements established by the *Traquair* case.

Nevertheless, other cases open the way to a broader approach towards forms of evidence. The French Supreme Court stated that, in this matter, the principle of freedom of evidence fully applies (Supreme Court, *Concordia I* case, 13/03/2007) and that photographs are not mandatory (Marseille High Court, *Cimil* case, 06/09/2004), the sovereign discretion of the judges being the fundamental principle.

5.3.2. *What answers and advances have been observed?*

While the reference used in the case of direct visual observation must necessarily be the Bonn Agreement Oil Appearance Code, the development of the use of remote sensing technology as well as certain recent decisions highlights certain advances. The *Carthage* case is of particular interest. The Aix-en-Provence Court of Appeal recognized (verdict on 08/02/2016) the reality of an operational discharge initially detected at night by SLAR, then confirmed by IR, together with the identification of the ship's position through AIS data, recorded in a regularly established report. There is therefore a shift towards a conviction based on "(...) the cross-comparison of different pieces of information from different sources". Mention can also be made of examples of advances observed, such as the decision by Truro Magistrates Court in the United Kingdom on October 4, 2013, in the *Maersk Kiera* case, in which evidence of the ship's position at the time of the discharge was ultimately provided by satellite imagery (EMSA operated satellite), thus confirming that the ship was in the territorial waters under the UK's sovereignty (condition required to allow legal action) [EMS 12].

5.4. Conclusion

The use of evidence obtained by remote sensing technologies, without direct visual observation, does not go against the principle of freedom of evidence presented above.

We know that the different technologies that exist can be used to detect operational discharge in ships' wakes, to characterize the type of pollutant and to provide the information required as concerns the materiality of the discharge, as well as the identity or position of the ship. Each remote sensing technique provides a share of the data that can be measured according to the technology used (temperature, concentration, etc.) without any risk of confusion. If direct observation is not possible, all the techniques and information available (SLAR, IR, UV, AIS data, satellite imagery) should therefore be considered together without exception, to corroborate the report and the observation made by the agent using the means available to him at that time.

To encourage court practices to evolve in this field, a reasonable approach would consist of raising magistrates' awareness of the existing technologies. In the future, the systematic inclusion of data obtained by remote sensing in reports of discharges detected in daylight, in addition to aerial photographs, would appear to contribute to this initiative.

Moreover, to consolidate the interpretation of this data and its probative nature, it would appear fundamental to have a reference comparable to the Bonn Agreement Oil Appearance Code but dedicated to remote sensing images. The aim is to enable the trial court to draw upon internationally recognized experimentation, which will serve as a reference for training agents in charge of drafting the report. The examination of this data during the trial hearing would be a useful complement to the report and the account of the agent, as has been the case of aerial photographs for over 20 years. The judicial handling of remote sensing evidence can only evolve through the practices of our specialized courts.

We can thus expect and indeed hope for an evolution of the methods of observing, collecting evidence of and judging cases of operational discharge given the reality of this form of pollution, which, as we know, has been adapted in response to our criminal policy.

5.5. References

[EMS 12] EMSA, Oil spill detection examples: Maersk Kiera, available at: http://www.emsa.europa.eu/csn-menu/use-cases/item/1873-satellite-images-as-primary-evidence-in-uk-court.html, European Maritime Safety Agency (accessed June 2020), February 25, 2012.

[PIN 17] PINHO G., VAGATA A., HENGSTERMANN T., "Optimized airborne oil spill remote sensing: poseidon, the quantitative approach", *Proceedings of the International Oil Spill Conference*, Long Beach, USA, pp. 1594–1611, May 15–18, 2017.

[ROY 06] ROY D., "Pollution en mer par rejets illicites d'hydrocarbures des navires: trois arrêts de la cour d'appel de Rennes", *Revue Juridique de L'environnement*, vol. 31, no. 3, pp. 265–287, 2006.

6

Long-Term Surveillance and Monitoring of Natural Events in Coastal Waters

6.1. Introduction

In the last decades, satellite observations have been largely used in operational oceanography [LET 15]. Satellite-derived mineral suspended particulate matter (SPM, related to turbidity and the light regime within the surface waters) and Chlorophyll-a (related to the abundance of phytoplankton) are major products obtained from the ocean color imagery [GOH 11]. At the European and global levels, these products are available through the CMEMS (Copernicus Marine Environment Monitoring Service). The CMEMS freely provides *in situ* data, model outputs and gridded products derived from satellite observations. These data are used for validating biogeochemical models and for long-term surveillance, as this is required by the European Marine Strategy Directive for estimating the eutrophication risk in coastal waters. They not only give unique information on the seasonal cycle of turbidity and phytoplankton at the scale of the continental shelf but also provide daily observations of exceptional events that may affect the ocean color, such as coccolithophore blooms, harmful algal plume, river plumes or resuspension of SPM after storms.

Chapter written by Francis GOHIN.

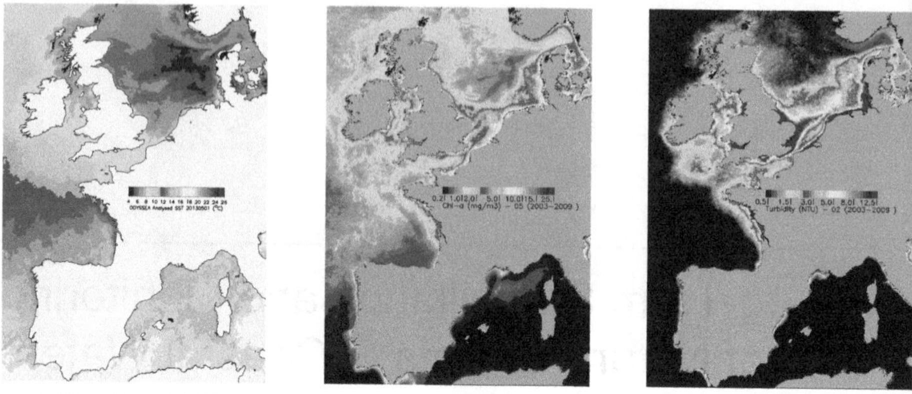

SST on May 1, 2013 (infrared data)

Chlorophyll-a in May (Ocean Color data)

Turbidity in February (Ocean Color data)

Figure 6.1. *Typical map for long-term surveillance. For a color version of the figure, see www.iste.co.uk/lefloch/remote.zip*

6.2. Satellite products for long-term surveillance

Sea surface temperature (SST) is derived from infrared sensors, while turbidity and Chlorophyll-a are derived from ocean color sensors (Figure 6.1). These quantities are closely related to climate variability and the evolution of nutrient fluxes in river plumes. The effect of the neap/spring tidal cycle and waves on the resuspension of sediment visible on the ocean color images has been studied in detail in Gohin *et al.* [GOH 15].

Figure 6.2 shows contrasted situations of SPM in March 2008 and March 2014. In March 2008, a late storm, which occurred during the spring tide, resulted in a considerable increase in SPM in the surface waters on March 12, whereas in 2014, the last storms highly affected the SPM at the beginning of March. Anticyclonic conditions after March 8, 2014 were followed by a fast decrease in the SPM concentration at the surface on March 12.

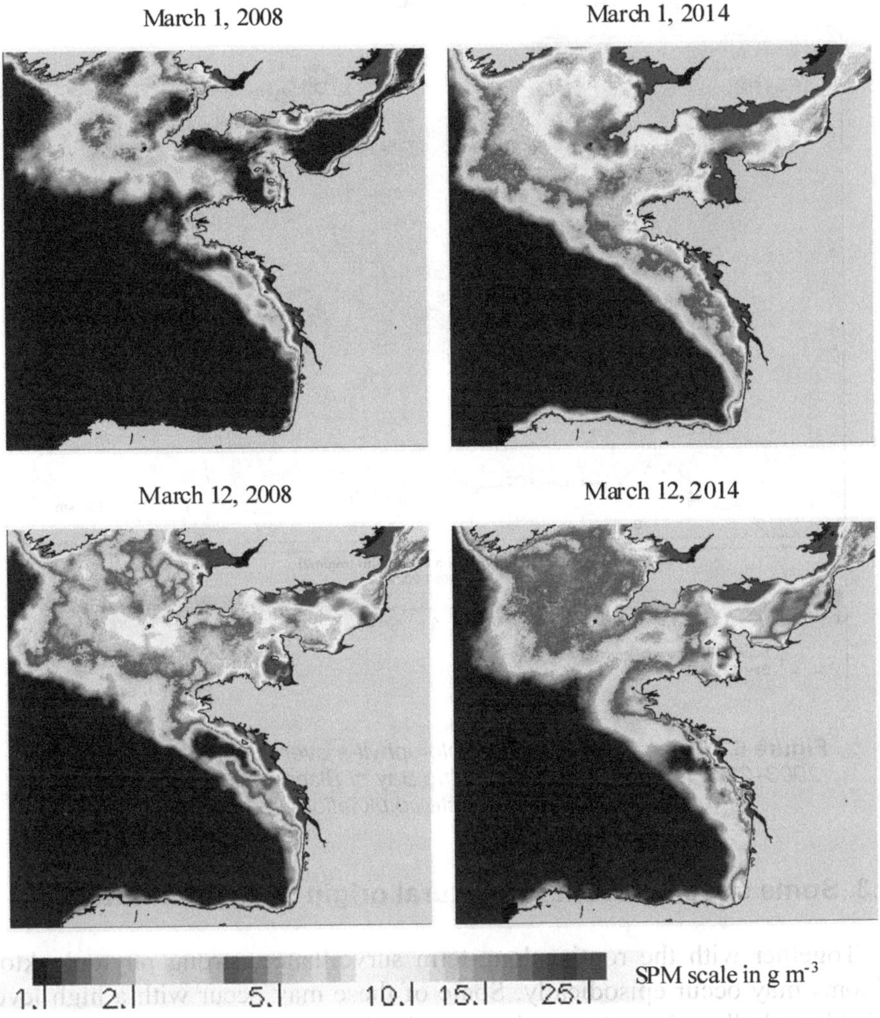

Figure 6.2. *Variability in the non-algal SPM at the beginning of March 2008 and 2014. For a color version of the figure, see www.iste.co.uk/lefloch/remote.zip*

Chlorophyll-a, an indicator of eutrophication risk in coastal waters, is also monitored from space. Figure 6.3 shows the percentile 90 of Chlorophyll-a over a six-year period from satellite and *in situ* data [GOH 10].

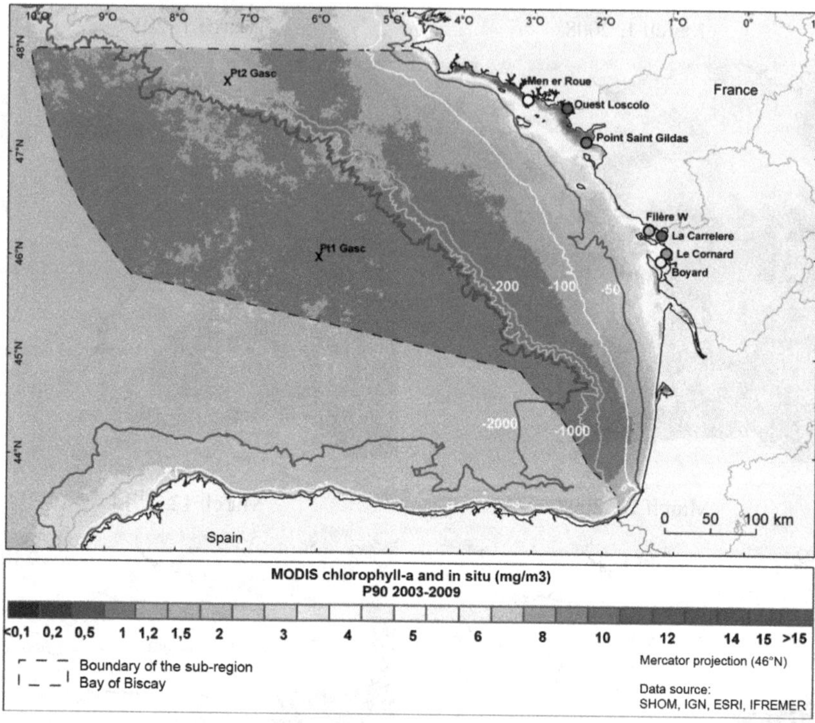

Figure 6.3. *The percentile 90 of Chlorophyll-a over the productive period 2003–2009 (March to October) in the Bay of Biscay. For a color version of the figure, see www.iste.co.uk/lefloch/remote.zip*

6.3. Some specific events of natural origin in coastal waters

Together with the routine long-term surveillance, strong phytoplankton blooms may occur episodically. Some of these may occur with a high level of chlorophyll with a strong signature in the satellite imagery. Figure 6.4 shows a harmful algal bloom of *Karenia mikimotoi* in the Western English Channel (WEC) in July 2003. These blooms often occur at the beginning of summer in the frontal region between the stratified waters of the Atlantic and the mixed waters of the WEC.

Blooms of coccolithophores, giving a whitish coloration to the surface waters, are also clearly visible from space from March to August, initially in the southern Bay of Biscay and appearing northwards throughout the spring and summer.

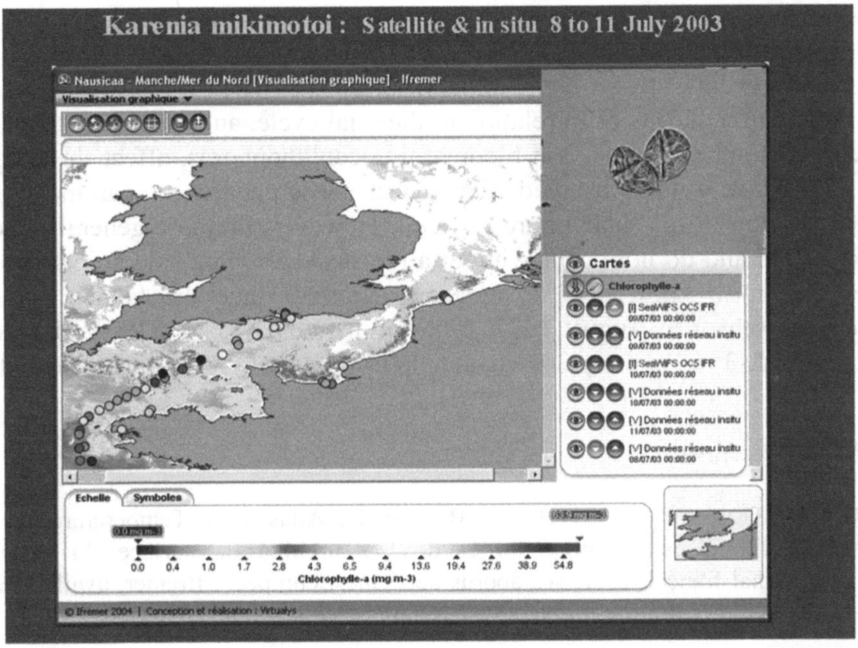

Figure 6.4. *A* Karenia mikimotoi *bloom observed* in situ *and from space in July 2003. For a color version of the figure, see www.iste.co.uk/lefloch/remote.zip*

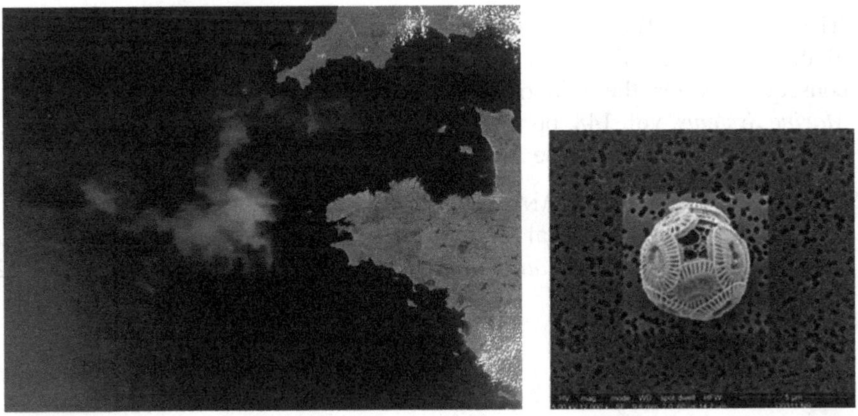

Figure 6.5. *A typical bloom of coccolithophore occurring in June on the margin of the continental shelf (MODIS image in RGB; source: NASA). An observation by electron microscopy of a cell of* Emiliania huxleyi *(source: IFREMER) is shown on the right*

6.4. Conclusion

Many events, from both mineral origins, such as resuspension of sediment after storms or in relation to the tidal cycle, and biological origins, such as strong phytoplankton blooms of coccolithophores, affect the colors of the surface waters observed from space. These phenomena can make the identification of oil slicks more difficult. However, they are generally well understood and do not occur without a physical or biological cause that we have learnt to know. The experience acquired since the launch of the SeaWiFS sensor in August 1997, followed by MODIS, MERIS, VIIRS and SENTINEL-3 in February 2016, can help us to identify most of these events.

6.5. References

[GOH 10] GOHIN F., SAULQUIN B., BRYERE P., Atlas de la Température, de la concentration en Chlorophylle et de la Turbidité de surface du plateau continental français et de ses abords de l'Ouest européen, Ifremer, available at: http://archimer.ifremer.fr/doc/00057/16840/ (accessed June 2020), 2010.

[GOH 11] GOHIN F., "Annual cycles of chlorophyll-*a*, non-algal suspended particulate matter, and turbidity observed from space and in-situ in coastal waters", *Ocean Science*, vol. 7, pp. 705–732, available at: http://www.ocean-sci.net/7/705/2011/os-7-705-2011.pdf (accessed June 2020), 2011.

[GOH 15] GOHIN F., BRYÈRE P., GRIFFITHS J.W., "The exceptional surface turbidity of the North-West European shelf seas during the stormy 2013–2014 winter: consequences for the initiation of the phytoplankton blooms?", *Journal of Marine Systems*, vol. 148, pp. 70–85, available at: http://archimer.ifremer.fr/doc/00252/36303/ (accessed June 2020), 2015.

[LET 15] LE TRAON P.-Y., ANTOINE D., BENTAMY A. *et al.*, "Use of satellite observations for operational oceanography: recent achievements and future prospects", *Journal of Operational Oceanography*, vol. 8, no. sup1, pp. s12–s27, 2015.

PART 3

Remote Sensing to Support the Response Strategy

7

VIGISAT Ground Receiving Station and EMSA CleanSeaNet Services

7.1. Introduction

The European Maritime Safety Agency (EMSA) has developed the CleanSeaNet service to support the EU member states in their activities to monitor marine oil spills caused by maritime traffic in the European EEZs. The CleanSeaNet service is designed to improve the efficiency of national inspections at sea, by delivering rapid oil spill alerts as soon as the radar satellite images reveal the presence of oil at sea. The member states have a responsibility to dispatch their marine patrol assets, verify the spills, and when possible, control polluter ships through application of the MARPOL convention which restricts the discharge of mineral oil at sea.

The timeliness of the data flow, from image acquisition to delivery in compact formats to the EMSA and the member states, is of the highest importance. Oil spill warnings are today delivered in less than 20 minutes. Further investigation such as detection of ships surrounding the spill [ARM 08] and identification by external means (AIS) is made within 30 minutes of image acquisition.

After eight years of operation, the CleanSeaNet has demonstrated its added value as an operational system, not only for oil spill surveillance, but also for surveillance of maritime traffic. Once integrated into a more powerful data management system such as IMDate, with appropriate data

Chapter written by Guillaume HAJDUCH.

fusion and behavior analysis functions, the observations from satellite radar images can also serve other applications. The CleanSeaNet service can now serve applications such as:

– detection of oil spill activities produced from ships in transit, the original CSN mission;

– detection of pollution produced by oil platforms, caused by accidental damages (explosion, wrong manipulation, storm) such as the Deepwater Horizon accident in the Gulf of Mexico;

– maritime activity surveillance in support of the European Fisheries Control Agency (EFCA) for detection of fishing vessels. The correlation of echoes with positions given by mandatory vessel monitoring systems should allow the detection of suspicious fishing vessels;

– maritime activity surveillance in support of Frontex for illegal transborder activities (immigration, smuggling, etc.).

For the past 15 years, several scientific organizations and operational services have used synthetic aperture radar (SAR) satellites to detect oil spills on the sea surface. The concept gradually evolved from R&D to pilot projects and operational systems. Oil spill detection now represents one of the most current applications of SAR satellites [PRA 11].

After the setup phase, the operations of the CleanSeaNet 2nd Generation started in February 2010 with ENVISAT segments. RADARSAT-1 and RADARSAT-2 segments were progressively incorporated in operations. Since the loss of ENVISAT in April 2012 and RADARSAT-1 in March 2013, the service has been based on the RADARSAT-2 mission. The third phase of the CleanSeaNet service ran from 2014 through to 2018. The service is mostly based on the European SAR satellites, the SENTINEL-1 mission (two spacecraft: SENTINEL-1A and SENTINEL-1B).

7.2. VIGISAT ground receiving station and detection of pollution in near-real time

CLS has provided reception and analysis of ENVISAT, RADARSAT-2 and SENTINEL-1 scenes throughout the CleanSeaNet service. The SAR scenes are received in VIGISAT in Brest, their analysis is performed by experienced personnel, and the reports are delivered to the EMSA with oil spill positions, date/time, oil spill dimensions and characterization, and metadata related to the sensor settings. The whole process can be performed within a maximum of 30 minutes. The oil spill reports are designed to be of immediate use for client Web-based browsers or in clients' software applications (typically to be ingested in a GIS or vessel traffic system).

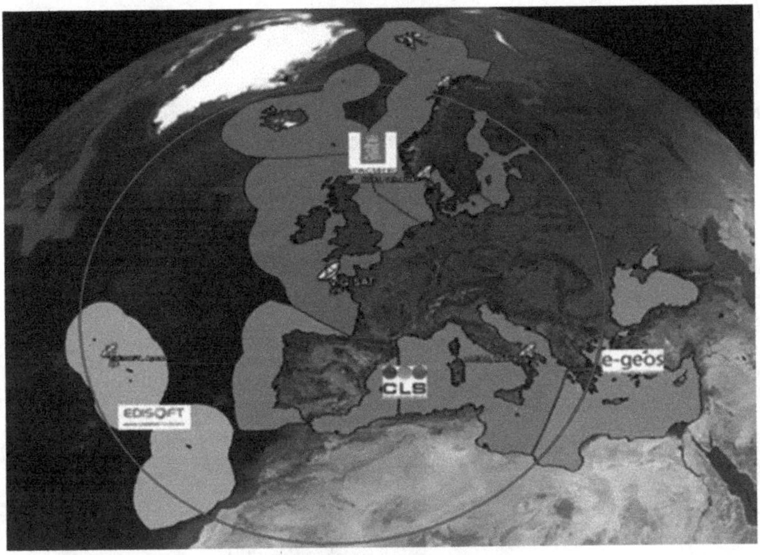

Figure 7.1. *Task areas of CleanSeaNet 3rd Generation. For a color version of the figure, see www.iste.co.uk/lefloch/remote.zip*

The echoes produced by ships in the image are also extracted as individual features (vessel detection reports). If resolution allows, the length and heading of the ship is measured.

From the same SAR satellite scene, sea surface roughness is processed and converted into a wind field. These products from the SAR scene are:

– oil spill detection reports;

– vessel detection reports;

– wind detection reports.

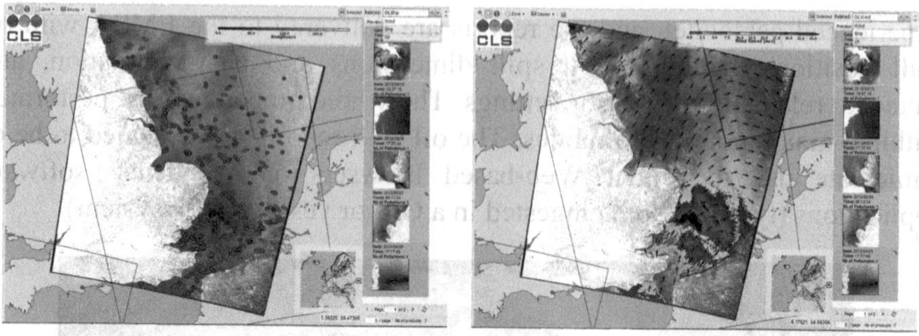

Figure 7.2. *Left: vessel detection reports with two classes of confidence. Right: SAR-derived wind fields. For a color version of the figure, see www.iste.co.uk/lefloch/remote.zip*

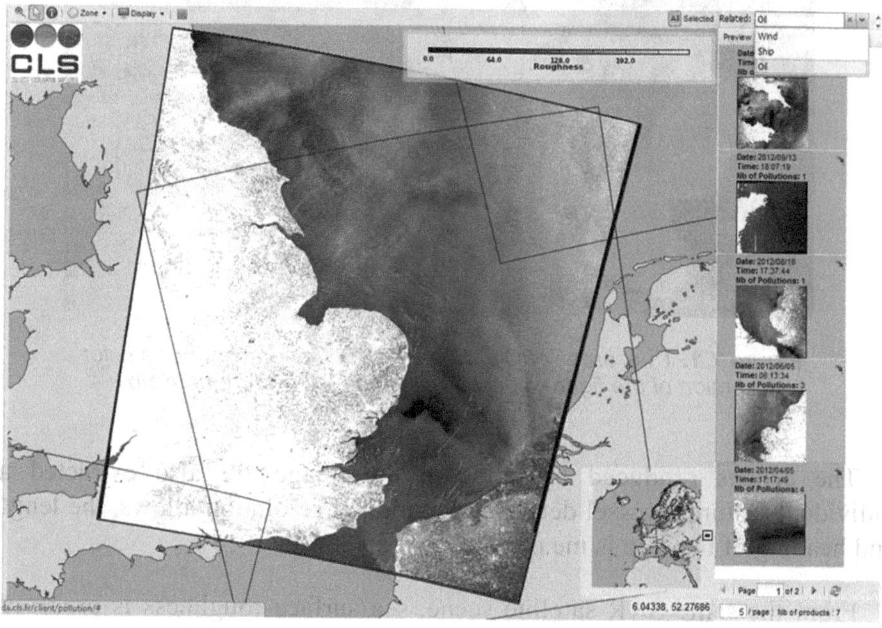

Figure 7.3. *Sea surface roughness and oil spill. For a color version of the figure, see www.iste.co.uk/lefloch/remote.zip*

The above products displayed in the CLS ©Maestro portal are all produced from the same scene:

– sea surface roughness and a linear oil spill contoured in orange (south east of England);

– vessel detection reports from the SAR scene (two classes);

– wind measurement from the SAR image.

7.3. Polluter identification with AIS data flows and drift modeling

A further evolution of the service includes the identification of polluters using AIS messages received from ships in the region of the spill. The AIS data are supplied by the EMSA from the SafeSeaNet system that gathers all AIS positions received along the European coastlines through terrestrial receiving antennas.

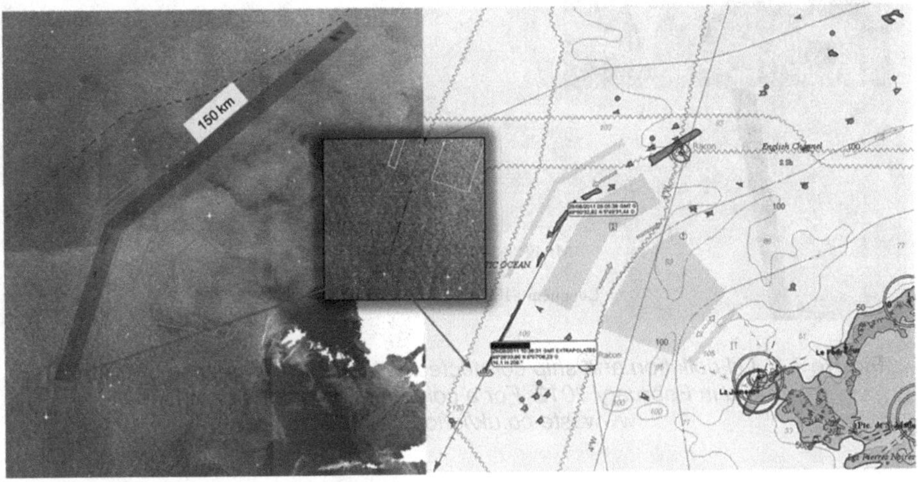

Figure 7.4. *Oil pollution and ship connected to the spill, with AIS identification from THEMIS GIS ENVISAT (©ESA). For a color version of the figure, see www.iste.co.uk/lefloch/remote.zip*

The methodology for oil spill detection and its association with a list of potential polluter identifiers is now mature. It was demonstrated in February 2012 during the MAERSK Kiera pollution report off the Scilly Islands [EMS 12]. The ship was detected causing pollution within the 12-nautical mile zone from the coast. The press based their reports on an investigation by the Maritime Coastguard Agency in charge of the EEZ. The satellite image was sufficiently explicit to make the shipping company admit its responsibility.

> "The unique trial at Truro Magistrates Court has been highlighted as the first-time satellite footage has been used by the Maritime and Coastguard Agency in the prosecution of a company for dumping waste illegally in the UK." [MAR 13]

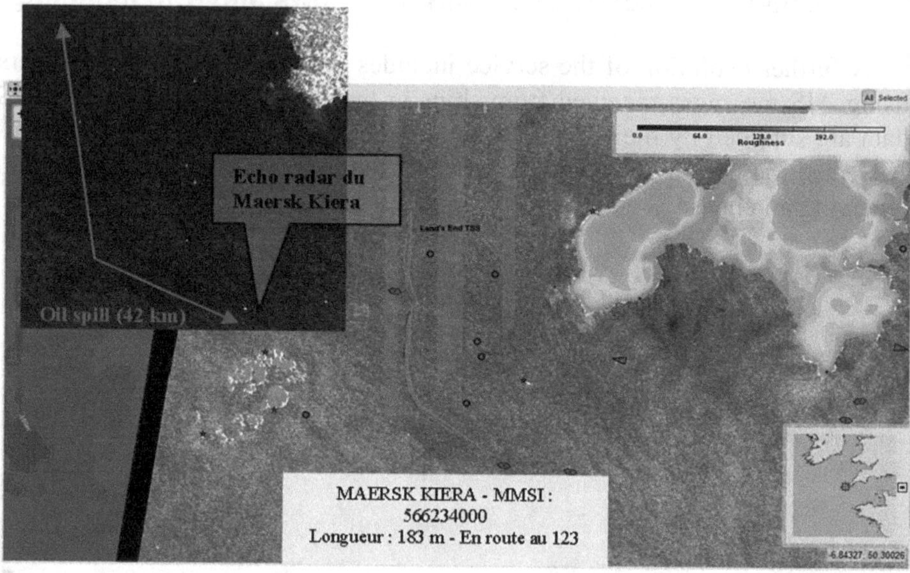

Figure 7.5. *Oil pollution and ship connected to the spill for the case of MAERSK Kiera in February 2012. For a color version of the figure, see www.iste.co.uk/lefloch/remote.zip*

Recently, CLS updated its operational service chain to integrate the drift component when analyzing AIS and detected oil spills. There are two possible ways to estimate the pollution source with drift modeling, the backward and forward approaches:

– when using the backward approach, the detected SAR-based oil spill is propagated in the past, and the pollution source is the target whose route intersects the backward propagated pollution in an optimal manner;

– in the forward approach, all the vessels are assumed to be a source of pollution: they pollute along their routes. The pollution source corresponds to the forward simulated pollution with the best agreement with the SAR-based detected oil spill.

The pros and cons of each methodology are outlined below:

– pros of the backward approach: this is the easiest approach in terms of algorithms and CPU time consumption as only the detected oil spill is drifted in the past;

– cons of the backward approach: the analyst must find the best spatial agreement while tuning the time scale in the past. No simple quantitative criteria can be implemented to assess the likely polluter. Analysis may be complex with spatial and temporal considerations;

– pros of the forward approach: the overall situation is "frozen" at one single time step, the SAR acquisition time. The analyst only performs a spatial analysis with distance–shape criteria between simulated and detected oil spills. Thus, it is possible to quantitatively assess whether the vessel is the likely polluter;

– cons of the forward approach: this is potentially time consuming since it implies the simulation of the drift for all the vessels potentially in the vicinity of the oil spill from its release until the SAR acquisition time. However, drift computation can be carried out before the SAR image acquisition, and as such is not a limiting factor for NRT operational services.

At the time of writing, CLS has successfully implemented the forward approach.

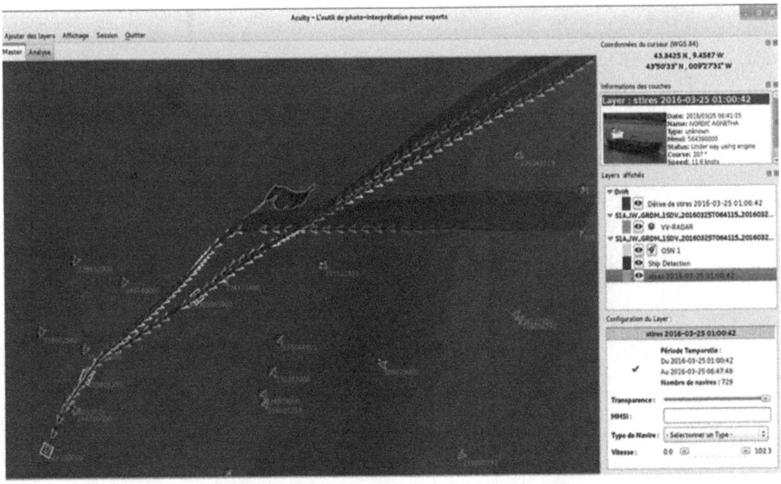

Figure 7.6. *Integration of drift modeling for polluter identification in the CLS ACUITY© platform. For a color version of the figure, see www.iste.co.uk/lefloch/remote.zip*

7.4. References

[ARM 08] ARMSTRONG L., SCHNEIDER H.A., FÄST O. *et al.*, "Integration of airborne AIS brings a new dimension to the detection of illegal discharge of oil spills", *Proceedings of the International Oil Spill Conference*, American Petroleum Institute, Savannah, USA, pp. 179–185, May 4–8, 2008.

[EMS 12] EMSA, Oil spill detection examples: Maersk Kiera, available at: http://www.emsa.europa.eu/csn-menu/use-cases/item/1873-satellite-images-as-primary-evidence-in-uk-court.html, European Maritime Safety Agency (accessed June 2020), 2012.

[MAR 13] MARITIME CONNECTOR, Tanker spied from space polluting seas off Cornwall, available at: http://maritime-connector.com/news/general/tanker-spied-from-space-polluting-seas-off-cornwall/ (accessed June 2020), October 10, 2013.

[PRA 11] PRAT L., LE BRAS J.Y., HAJDUCH G., "Oil spill detection based on spaceborne sar technologies: on-going services and perspectives", *Proceedings of the International Oil Spill Conference*, American Petroleum Institute, Portland, Oregon, USA, May 23–26, 2011.

8

System-to-system Interface Between the EMSA CleanSeaNet Service and OSERIT

8.1. Introduction

The European Maritime Safety Agency (EMSA) and the Royal Belgian Institute of Natural Sciences (RBINS) develop and operate together a system-to-system interface between the EMSA's CleanSeaNet service and OSERIT, the Belgian Oil Spill Evaluation and Response Integrated Tool. This interface is meant to provide CleanSeaNet users with a support tool for early and automatic oil drift and fate simulation results of any satellite-detected oil spills reported by the CleanSeaNet service in the North Sea and the English Channel. In view of the automatic forecast and backtrack simulations results, CleanSeaNet users have the possibility of further refining this early risk assessment either by activating their own national decision support system or by requesting new, advanced simulations through the CleanSeaNet GIS viewer.

This interface is currently passing the final acceptance tests. However, the system has already been used by RBINS for the oil pollution event subsequent to the *Flinterstar* sinking at 8 km off the port of Zeebruges on October 6, 2015. This event perfectly illustrates the potential synergies of remote sensing and modeling in cases of marine pollution and their integration in risk assessments that must be performed for any significant pollution of the marine system.

Chapter written by Sébastien LEGRAND and Ronny SCHALLIER.

Marine pollution is a global environmental problem. If large accidental oil pollution and its dramatic environmental impacts receive significant public attention, the marine environment also suffers from chronic operational oil pollution discharge of oil sludge, oily ballast from fuel tanks, machinery space bilges and oil cargo residues. To protect the marine environment from these discharges, a series of increasingly stringent regulations have been being successfully implemented from the 1950s onwards at global (OILPOL 1954/1962, MARPOL 73/78, etc.), continental (EU directives 2000/59/EC, Directive 2008/56/EC, etc.), regional (e.g. Bonn Agreement 1969/1983/1989, etc.) and national levels [LAG 12].

Surveillance plays a key role in this global strategy of marine pollution reduction. For instance, the Bonn Agreement was amended in 1989, committing North Sea Coastal States to carry out regular surveillance activities in their zone of responsibility. At the national level, this surveillance primarily relies on airborne platforms (airplanes, helicopters and drones) on which can be mounted various remote sensors such as trained operators' eyes, side-looking airborne radar (SLAR), visible-band cameras, infrared-band cameras or other night vision devices and in some cases multi-spectral or hyper-spectral cameras. In addition, since the mid-2000s, airborne surveillance programs are completed with satellite-based synthetic aperture radar (SAR) surveillance services such as the EMSA CleanSeaNet service.

Airborne and satellite-based remote sensing are complementary tools, each having their pros and cons:

– the pixel brightness of a radar image (SLAR or SAR) is actually a measurement of the local sea surface radar backscattering. Since surface slicks significantly dampen the amplitude of the wind-generated short gravity-capillary waves, they appear on radar images as well-contrasted and well-defined dark shapes. However, many other phenomena can decrease the sea surface radar backscattering so that any radar detection should be confirmed by a direct observation. If SLAR detections can easily be confirmed by the operators of airborne surveillance flights, false-positive detections remain a problematic issue for satellite-based surveillance;

– to some extent, a trained operator of airborne surveillance flights can estimate pieces of information describing a marine pollution such as the probable type of contaminant, the slick thickness, the spilt volume or the weathering state. Although radar images provide accurate information on the position and extension of a potential slick, the above-mentioned characteristics cannot be retrieved from radar images;

– airborne surveillance is extremely flexible. In a couple of hours, a surveillance flight can be activated. Its flight plan can also be adjusted according to observations. Satellite surveillance is by far less flexible as it is constrained by the satellite orbits;

– satellite-based SAR images cover a very large area as they are acquired in segments of up to 1,400 km and swaths of up to 500 km;

– while airborne surveillance flights can be canceled in case of strong wind gusts, low cloud ceiling or bad visibility, SAR satellite imagery is almost unaffected by bad weather conditions. On the contrary, since a minimal wind speed of 2–3 m/s is needed to get enough brightness and contrast in the radar images, overly calm weather conditions negatively affect satellite-based surveillance.

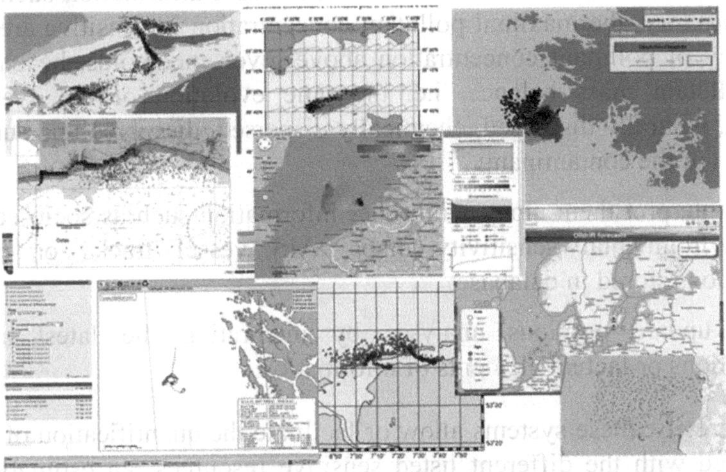

Figure 8.1. *Snapshots of the main decision support systems used in the North Sea by public authorities, including OSERIT, Oilmap, SeaTrackWeb, Mothy, OD3D, BSHmod.L and OilTrans. See [BON 16] for the most recent exhaustive overview of the decision support systems used in cases of marine pollution (image courtesy: RBINS, met.no, Meteo-France, BSH, FCOO, Deltares, Marine Institute and CEFAS, 2013). For a color version of the figure, see www.iste.co.uk/lefloch/remote.zip*

The airborne and remote sensing surveillance programs naturally aim to detect marine pollution and monitor its evolution. They mainly provide snapshots of a pollution event and must therefore be further analyzed and interpreted by coastguard operators or other national responsible authorities to get a full and comprehensive understanding of the pollution and its consequences. In particular, in cases of significant pollution, one of the priorities is to assess the risk posed by this pollution (1) to the public and responder safety, (2) to the marine environment and (3) to the socioeconomic assets on which a state or coastal community depend.

While 10 years ago these assessments were mainly an "expert judgment matter", they presently increasingly rely on decision support systems that integrate at least a spill drift and fate model (Figure 8.1). The most advanced systems allow their operators within a few clicks:

– to describe the pollution;

– to perform forward or backward spill drift and fate simulations according to the most recent and reliable met-ocean forecasts;

– to access and visualize the resulting relevant information such as spill drift trajectory, the maximal pollutant concentration in sensitive areas, time residency of pollutant concentration above given thresholds, beaching risk and pollutant mass balance, i.e. the time evolution of the evaporated, floating, naturally dispersed, chemically dispersed, dissolved and sunk mass fractions of the contaminant;

– to interpret them along with other information such as socio-economic and environmental sensitivity maps, AIS vessel tracks or pollutant descriptions stored in database;

– to update previous analyses by integrating the latest available observations or factual elements of information.

Concretely, these systems allow or facilitate the quantification of the risk associated with the different listed sensitive resources; an estimation – as objectively as possible – of the efficiency of the different possible response strategies; the guidance of surveillance teams towards spills observed a few hours ago; the identification of potential polluters; the delimitation of safety and/or exclusion areas and the design of the (post-)crisis environmental monitoring program. In addition, their graphical outputs are increasingly used by crisis managers for communication purposes.

In order to provide faster and easier access to information for operational decision-making for oil spill surveillance, potential polluter identification and spill response, the European Maritime Safety Agency (EMSA) implements a system-to-system interface between the CleanSeaNet service and several leading European oil spill drift and fate model service providers. Two interfaces are being implemented for the Baltic Sea and the North Sea, respectively, while two other demonstration cases are envisaged for the Portuguese Atlantic Region and the Mediterranean Sea.

The purpose of this article is to present the North Sea region-wide implementation of the system-to-system interface between the EMSA CleanSeaNet service and the Belgian operational oil spill drift and fate model OSERIT. The article is organized as follows. First, both the CleanSeaNet and the OSERIT systems are briefly described. Then, the philosophy of the system-to-system interface is explained. Finally, the potential synergies between remote sensing and modeling is illustrated using the oil pollution event subsequent to the *Flinterstar* sinking at approximately 8 km off the port of Zeebruges on October 6, 2015.

8.2. The EMSA CleanSeaNet service

EMSA helps to identify, trace and track illegal oil discharges and polluters through the satellite image-based service CleanSeaNet. Synthetic aperture radar images are used to detect the position and extension of potential oil slicks on the sea surface and adjacent vessels. This monitoring service supports coastal states and the European Commission and offers a solid basis on which to expand illegal discharge-related measures and to support response operations to accidental pollution.

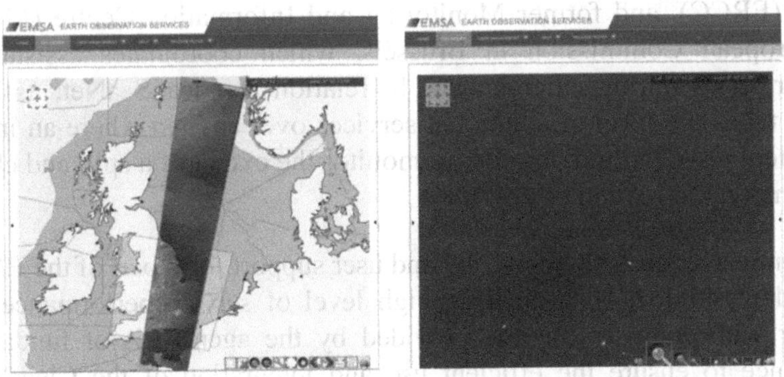

Figure 8.2. *Example of a SAR image made available to CleanSeaNet users through the CleanSeaNet GIS viewer and a close-up view of potential spill detections at 20 km off the British coastline. For a color version of the figure, see www.iste.co.uk/lefloch/remote.zip*

In addition to RADARSAT-2, the synthetic aperture radar satellites and sensors now in use include SENTINEL-1 and the TerraSar-X and Tandem-X constellations. Optical imagery from a wide range of satellites is also available. Coastal states define their service coverage requirements. In cooperation with the users, EMSA plans and orders satellite images to meet these requirements. Accordingly, satellite data are acquired via a network of receiving stations.

Operators assess the images, together with supporting meteorological, oceanographic and ancillary information (AIS, vessel detection) where available, to identify possible polluters, to determine the likelihood of the presence of oil on the sea surface and to assist in identifying the source of the pollution. If a potential spill is detected (Figure 8.3), it is of utmost importance that coastal state administrations are immediately alerted by phone and email with the aim of increasing the likelihood of catching a polluter red-handed.

As it is critical for confirming a possible spill and catching polluters in the act, the shortest possible delay between satellite detection and alert is essential for a rapid response by coastal states. CleanSeaNet is a near real-time service. Detection results are reported to the affected coastal state approximately 30 minutes after satellite image acquisition (the exact time varies according to the size of the image).

In addition to the regular monitoring service, the agency also provides assistance to member states during emergency situations. This is usually requested by member states through the Emergency Response Coordination Centre (ERCC), and former Monitoring and Information Centre (MIC), of the European Commission in Brussels, which coordinates assistance to member states during emergencies. In relation to CleanSeaNet, assistance usually takes the form of additional services over an area where an incident or accident has occurred, in order to monitor the extent of a spill and changes over time (e.g. direction of drift).

CleanSeaNet quality monitoring and user support form part of the EMSA's day-to-day activities to keep up a high level of service performance. This, together with end-user training provided by the agency, is of fundamental importance to ensure the efficient use and integration of the CleanSeaNet service into national operations.

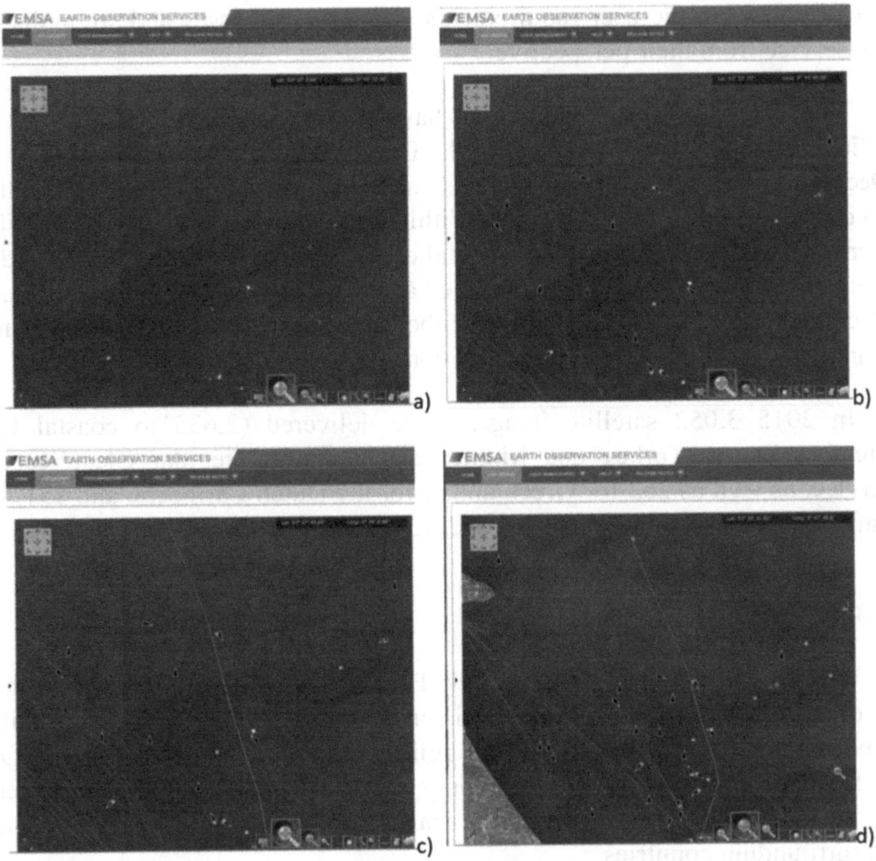

Figure 8.3. *In cases where possible slicks are detected, coastal state administrations receive a report within 30 minutes from the SAR image acquisition time. In order to help catch a polluter red-handed, the coastal state operator can log into the CleanSeaNet Web-based GIS viewer in order to better visualize the slicks (a) and all the vessels present in the scene as well as their AIS track if available (b). On the basis of these observations, the operator can identify a subset of potential polluters (c) who are not necessarily the vessels located closest to the slicks (d). For a color version of the figure, see www.iste.co.uk/lefloch/remote.zip*

As CleanSeaNet is linked to national and regional response chains, interaction with user communities in the coastal states is important for ensuring the best use of the service and identifying improvements to be made. To that effect, CleanSeaNet User Group meetings are organized regularly by EMSA. These events are an opportunity to foster better communication with authorities using CleanSeaNet in the coastal states and get feedback on the user requirements and on weaknesses and strengths of the service from their perspective.

The first three years of operations have demonstrated that CleanSeaNet is efficient for the detection of oil spills. During the period April 16, 2007, to December 31, 2009, 5,816 images were successfully delivered to the 26 coastal states using the service. Within these images, 7,193 possible spills were detected by CleanSeaNet. A total of 1,997 were verified on site by the member states and 542 were confirmed as being mineral oil. The overall rate of confirmation is better than 50% if the spill is checked by aircraft no later than three hours after satellite acquisition.

In 2015 3,052 satellite images were delivered (2,635 to coastal EU Member States, Turkey and Montenegro under the regular CleanSeaNet service, as well as 262 for Greenland, 4 for the Dutch Caribbean and 151 for Safemed and Traceca) and a total of 2,423 possible oil spills were detected.

8.3. OSERIT

OSERIT is the Belgian Oil Spill Evaluation and Response Integrated Tool, a 24/7 accessible support decision system for evaluation of oil spills reported in the North Sea and the English Channel. It is made of a 3D Oil drift and fate model and a user-friendly Web-based decision support system restricted to registered users, mainly coast guard operators from Belgium and its surrounding countries.

8.3.1. *The OSERIT Oil Spill Model*

The OSERIT model has been designed to balance the state of the art in oil spill modeling with the reality of the field and operational needs, i.e. mainly the short time window for intervention and the common lack of information on the spill itself.

The OSERIT model is a kinetic model based on empirical data and parameterizations that simulates the 3D drift and fate of oil at the sea surface and in the water column using the usual Lagrangian particle tracking approach (Figure 8.4).

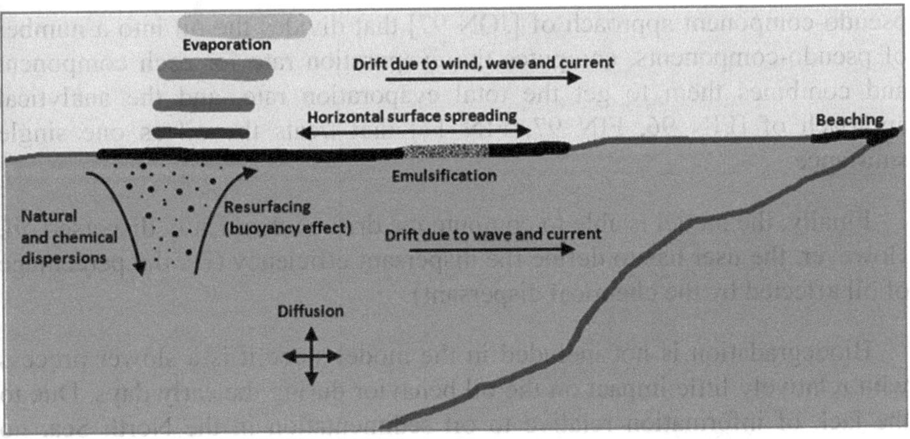

Figure 8.4. *Physical processes included in the OSERIT oil spill model*

Advection at the sea surface is computed as a result of the combined effect of winds (leeway drift), surface water currents and waves (Stokes drift). Within the water column, particles move under the action of water currents, waves and the buoyancy effect. The horizontal and vertical turbulent diffusive transport is expressed using the random walk technique following [WAN 08]. Particles can move from the surface into the water column through the process of vertical natural dispersion. Surface oil is then split into smaller droplets that are propelled into the water column. The oil natural dispersion follows the kinetic approach of [TKA 02], but its rate can also be defined by the user. Two different parameterizations have been implemented in OSERIT to simulate the horizontal surface spreading of oil over the water surface mainly due to the combined effect of gravity, inertia, interfacial tension and viscosity. The first method is based on [FAY 71] and approaches the spreading due to gravity-viscous forces by computing random velocities in the range of velocities that are assumed to be proportional to the diffusion coefficients [GAR 99]. The second method, also based on [FAY 71], considers the oil slick as an ellipse that elongates along the wind direction [LEH 84]. The surface horizontal spreading stops when the terminal thickness is reached based on data from [MCA 87], as in [FRE 03]. Beaching is included.

Oil weathering processes and their effects on oil characteristics can also be computed by the model. Emulsification is computed as a function of oil characteristics and significant wave height as in [SCO 05]. Oil evaporation can be computed following one of the two following approaches: the pseudo-component approach of [JON 97] that divides the oil into a number of pseudo-components, computes the evaporation rate for each component and combines them to get the total evaporation rate, and the analytical approach of [FIN 96, FIN 97, FIN 11] that treats the oil as one single substance.

Finally, the model is able to compute the drift of chemically dispersed oil. However, the user has to define the dispersant efficiency (i.e. the percentage of oil affected by the chemical dispersant).

Biodegradation is not included in the model since it is a slower process with relatively little impact on the oil behavior during the early days. Due to the lack of information relative to oil sedimentation in the North Sea, no such parameterization has been implemented yet in the OSERIT model.

The model can be run forward in time to provide a forecast of the oil spill drift and fate or backward in time to provide a backtracking of oil spill. For maximum flexibility, each process implemented in the model can be activated (or deactivated) by designated model switches. The complete description of the model equation can be found in [DUL 13].

It is important to note that oil slicks drift away under the influence of winds, waves and ocean currents. As a result, the accuracy of the simulation depends, to a large extent, on the reliability of the wind, waves and current data used. The oil drift model makes it possible to perform a large range of simulations; from 2D drift-only simulations to 3D simulations that include horizontal surface spreading, natural vertical dispersion, buoyancy and weathering. Some of these last processes strongly depend on oil type. As a result, the simulation accuracy relies on oil characteristics given by the user.

8.3.2. *OSERIT visualization tool*

To enable 24/7 accessibility by all Belgian Coast Guard Agencies, a Web-based interface has been developed [OSE 20]. This interface was designed to be as user-friendly as possible. It offers two main functionalities: a simulation manager and a visualization tool.

The simulation manager allows users to manage existing model simulations and to request new ones. To this end, they must define the simulation setup by filling in a setup form. Basic inputs consist of initial time and location of the slick and simulation type (forward or backward in time). Advanced options can include oil characteristics, oil release and spill scenarios, physical oil processes to be taken into account and the possible use of chemical dispersant.

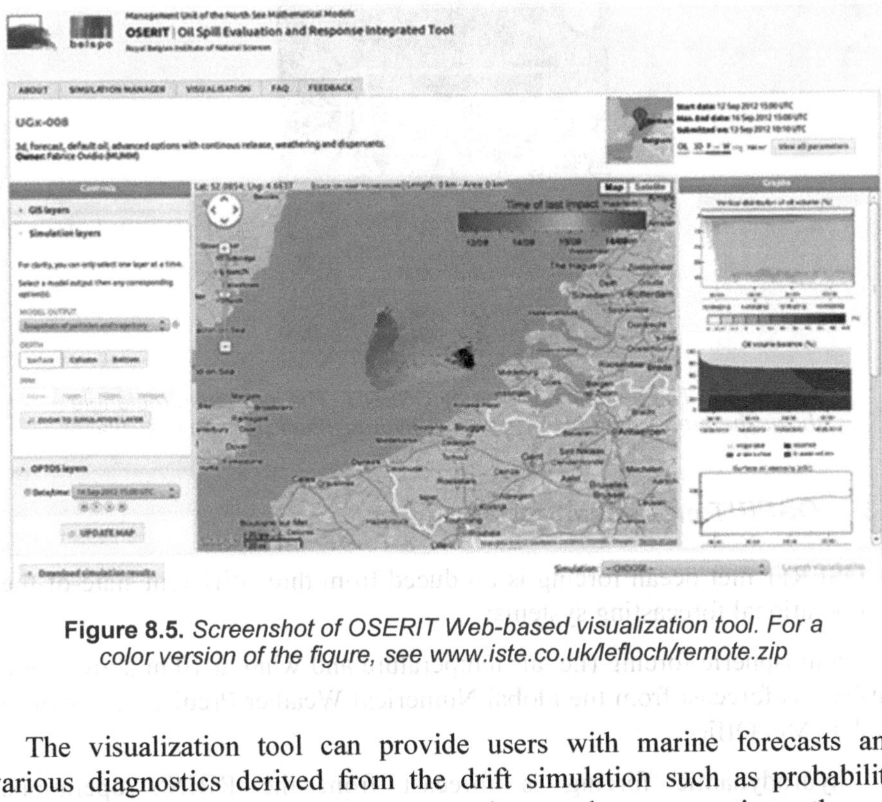

Figure 8.5. *Screenshot of OSERIT Web-based visualization tool. For a color version of the figure, see www.iste.co.uk/lefloch/remote.zip*

The visualization tool can provide users with marine forecasts and various diagnostics derived from the drift simulation such as probability maps, oil-impacted areas, oil concentrations and exposure times through geographical overlays. Part of the interface is based on Google Maps and makes it possible to superimpose model results with other environmental and economic GIS layers. Time series of oil vertical distribution, volume balance and oil characteristics can also be displayed. Results are available for download from the visualization tool (Figure 8.5).

8.3.3. *OSERIT domain*

The domain of the OSERIT model is presented in Figure 8.6. It covers the English Channel from 4°W and the southern North Sea up to 57°N. During model simulations, any oil that is pushed out of the domain covered by OSERIT is considered as "lost" and can therefore not come back in. This is a problem for spill that can occur close to the OSERIT boundaries.

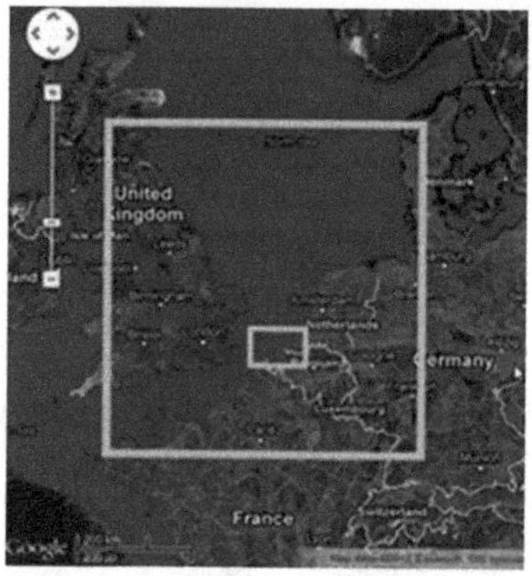

Figure 8.6. *The domain covered by OSERIT. The inner blue rectangle shows the area where high-resolution hydrodynamic forcing is available*

8.3.4. *OSERIT met-ocean forcing*

OSERIT met-ocean forcing is produced from three different state-of-the-art operational forecasting systems:

– atmospheric forcing (i.e. air temperature and wind at 10 m above water surface) is forecast from the global Numerical Weather Prediction system of the UK Met Office;

– hydrodynamic forcing is forecast from the RBINS operational hydrodynamic models OPTOS-NOS and OPTOS-BCZ (including 3D current, sea surface elevation and turbulent vertical diffusivity). Their horizontal resolutions are of 5 km and 750 m, respectively;

– sea state (including wave period, direction and significant height) is forecast from the RBINS operational implementation of the wave modeling (WAM) model.

8.3.5. *OSERIT oil database*

Some oil processes implemented in the OSERIT model (such as evaporation, horizontal surface spreading, natural vertical dispersion, emulsification and buoyancy) strongly depend on oil properties. It is therefore crucial to provide the right oil characteristics to the model. To help with the lack of information on oil characteristics in most cases of marine pollution by oil, the OSERIT model was built with an oil database. This database was kept intentionally short for ease of use and includes a wide range of oil types that are likely to cross the Southern part of the North Sea (Figure 8.7). Oil types go from light, medium, heavy and very heavy crude to main fuel types. Characteristics of oil involved in the *Erika* and the *Prestige* incidents are also available. The complete list of available oils is given in Figure 8.8. The OSERIT database is based on oil characteristics provided in ADIOS, [JOK 99] and [FIN 11] and by CEDRE.

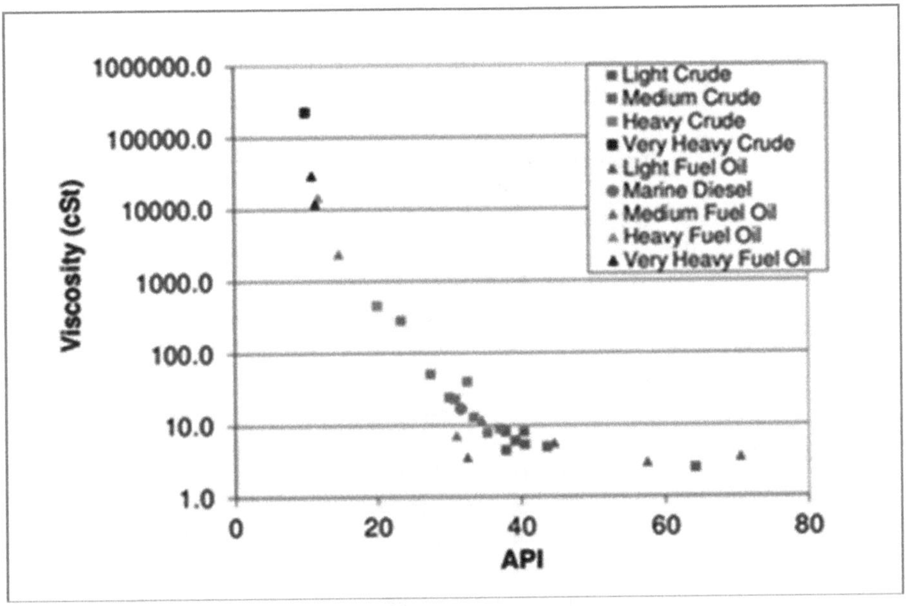

Figure 8.7. *Scatter plot between viscosity and API for each oil type of the OSERIT database. Viscosity is given in cSt at 15°C [DUL 13]. For a color version of the figure, see www.iste.co.uk/lefloch/remote.zip*

Oil type selection

- Light crude
  - Brent blend
  - Algerian condensate
  - Brega
  - Ekofisk
  - Forties blend
  - Saharan blend
  - Statjord
- Medium crude
  - Arabian
  - Arabian extra light
  - Arabian heavy
  - Arabian light
  - Arabian medium
  - Iranian heavy
  - Alamo
  - Oseberg
- Heavy crude
  - Alba
- Very heavy crude
  - Boscan
- Light fuel oil
  - Unleaded Interm Gasoline
  - Fuel oil N°1 (Avjet A)
  - Gasoline blending stock (Alkylate)
  - Gasoline blending stock (Reformate)
- Marine diesel
  - Marine diesel
- Medium fuel oil
  - Fuel oil N°2 (diesel)
  - Fuel oil N°2 (Ho/diesel)
  - Intermediate fuel oil 180
- Heavy fuel oil
  - Intermediate fuel oil 360
- Very heavy fuel oil
  - Erika
  - Prestige

Figure 8.8. *Oil type list available within the OSERIT oil database [DUL 13]*

8.4. A system-to-system interface between CleanSeaNet and OSERIT

Since oil spill modeling provides valuable information which could assist in the identification of vessels responsible for the illegal discharge (spill backtracking) and in the prediction of spill drift and fate (spill forecasting) to support decision-making for pollution response activities, EMSA has implemented a generic system-to-system interface between the CleanSeaNet Data Centre and a generic "Model Service Provider" in order to perform the following actions in a fluent and fully automatized way:

– submission of a simulation request package to the model service provider through a dedicated SFTP box and/or an OGC Web Processing Service;

– execution of the simulation request by the model service provider;

– uploading in a dedicated SFTP box of an output package that includes the simulation results in a file format that can be directly ingested in the CleanSeaNet Data Centre;

– visualization of the simulation results in the CleanSeaNet GIS viewer.

The request package is made of:

– one "package information XML file" that describes the content of the package;

– one "Oil Spill feature GML file", i.e. a file format routinely used by the CleanSeaNet Data Centre to store the characteristic of any possible oil spill detections. This includes the geometry of the spill (a polygon that describes the external contour of the slick, as well as data about its length, width and area) and ancillary data such as the CleanSeaNet spill identifier, the scene acquisition time and, if available, the oil type, the weathering status (in days adrift), the oil thickness, viscosity and density. These last five elements of information are only available for the confirmed spills;

– one "process request XML file", i.e. a file that fully describes the requested model setup. This includes at least the start time of the simulation and its duration; the number of Lagrangian particles, the output time step and the simulation type (forward or backward in time). In specific cases, the file can also include or supersede the exact oil type, thickness, density and viscosity, total spill volume and discharge duration.

The output package is made of:

– one "package information XML file" that describes the content of the package;

– one "model output description file" that provides a model output summary, including the trajectory of the center of mass of the spill;

– a series of "Oil Spill feature GML files" (one file per model output time step) that can be directly inserted in the CleanSeaNet Data Centre to describe the simulated time evolution of the surface slick characteristics: external contour(s) and surface oil thickness, viscosity and density;

– optionally, a "netCFD file" describing the oil concentration in the water column.

EMSA and RBINS have jointly implemented, tested and fine-tuned this system-to-system interface between CleanSeaNet and OSERIT for the North Sea and the English Channel. Four different use-case scenarios have been identified:

– automatically triggered forecast and backtrack scenarios that are generic scenarios to be applied to every satellite-detected oil slicks reported by CleanSeaNet on the OSERIT domain;

– manually triggered forecast and backtrack scenarios that allow CleanSeaNet users to refine the results provided by the automatically triggered scenarios.

8.4.1. *Scenario 1: automatically triggered forecast*

The automatically triggered forecast scenario consists of an instantaneous release of 1,000 Lagrangian particles that are randomly scattered at the sea surface within the given closed polygon provided in the "process request package". The 3D drift simulation takes into account the drift due to current, wind and waves (stokes drift), oil endogenous spreading at the sea surface, natural vertical dispersion and horizontal turbulent diffusion if the oil Lagrangian particle is within the water column.

Due to the fact that no reliable information on the oil type and initial oil weathering is available from the satellite image, oil weathering is not computed in this scenario (i.e. no oil evaporation, emulsification and time evolution of the oil density and oil viscosity at the sea surface). However, oil spreading and natural dispersion are computed assuming that the spilt oil closely behaves like a fresh marine diesel.

The simulation length is set to 72 hours.

8.4.2. *Scenario 2: automatically triggered backtrack*

The automatically triggered backtrack scenario consists of an instantaneous release of 1,000 Lagrangian particles that are randomly scattered at the sea surface within the given closed polygon provided in the "process request package". The 2D surface backtracking takes into account the drift due to current, wind and waves as well as horizontal diffusion at the sea surface. The horizontal diffusion aims to provide some confidence interval for identifying potential polluters.

In this automatically triggered backtrack scenario, oil spreading, oil weathering and natural vertical dispersion and resurfacing processes cannot be computed.

The backtrack simulation length is set to 24 hours.

8.4.3. *Scenario 3: manually triggered forecast*

The manually triggered forecast scenario aims to offer CleanSeaNet users the possibility of refining the automatically triggered forecast scenario.

In comparison with the automatically triggered forecast scenario, the CleanSeaNet users are able to:

– specify the number of Lagrangian particles to be used (with a maximum of 2,500 particles);

– specify the starting time, the duration of the leak, the depth of the release and the oil discharge in case of a continuous release;

– specify the oil type;

– supersede the initial oil viscosity [in cSt at 15°C] and density [in kg/m^3 at 15°C];

– compute the oil weathering with the limitation that the initial weathering state is fresh oil.

8.4.4. *Scenario 4: manually triggered backtrack*

The manually triggered backtrack scenario aims to offer CleanSeaNet users the possibility of refining the automatically triggered backtrack scenario, taking into account the possibilities and limitations offered by the "process request file". Combined with the fact that OSERIT limits backtracking simulations to 2D surface drift, this means:

– specifying the number of Lagrangian particles to be used (with a maximum of 2,500 particles);

– specifying the starting time, the duration of the leak, the depth of the release and the oil discharge in case of a continuous release;

– performance of the system-to-system interface.

The interface is currently passing its final acceptance tests. The tests show that the automatically triggered forecast and backtrack simulation results are made available in the CleanSeaNet GIS viewer within 15 minutes from the spill alert (Figure 8.9).

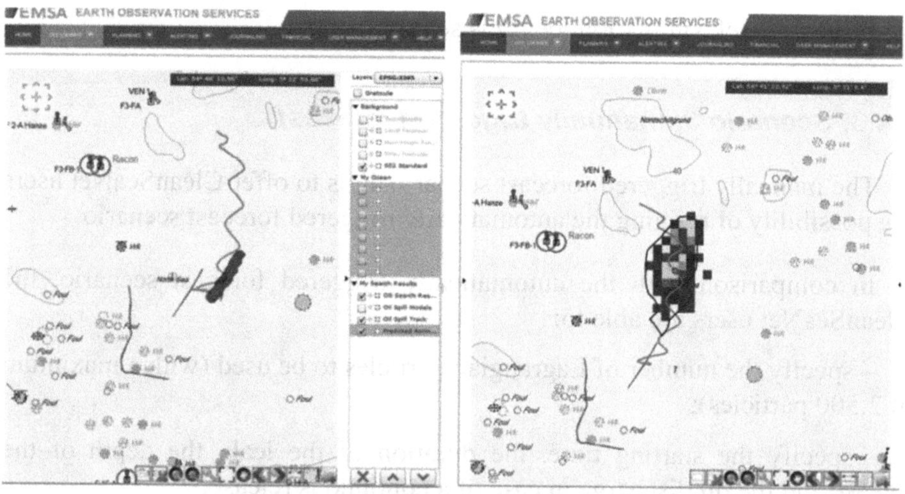

Figure 8.9. *OSERIT forecast simulation results as presented in the CleanSeaNet GIS viewer 15 minutes after the spill alert. The detected oil spill (in red, left and right), the trajectory of the center of mass of spill (in red, left and right), the snapshot of the slick contour (in blue, left) and the snapshot of the oil concentration (multicolor, right) are shown. A control panel (not shown) allows easy navigation in time. For a color version of the figure, see www.iste.co.uk/lefloch/remote.zip*

8.5. The *Flinterstar* incident

8.5.1. *The incident*

At 04:10 CET on October 6, 2015, the LNG tanker *Al Oraiq* collided with the general cargo vessel *Flinterstar* in the coastal shipping lane Scheur located in the Belgian Territorial Sea near the port of Zeebruges (Figure 8.10). The *Al Oraiq* remained stable, suffering limited hull damage at its bow, and did not release any oil or other harmful substances; the tanker soon sailed to the port of Zeebruges for inspection. The *Flinterstar* however was severely damaged in its mid-section, and the vessel was pushed onto a very shallow (−10 m) coastal sandbank named Wandelaar, where it sank as a "total loss" at position 51°24.007'N – 003°03.717'E, at merely 8 km NW of the port of Zeebruges.

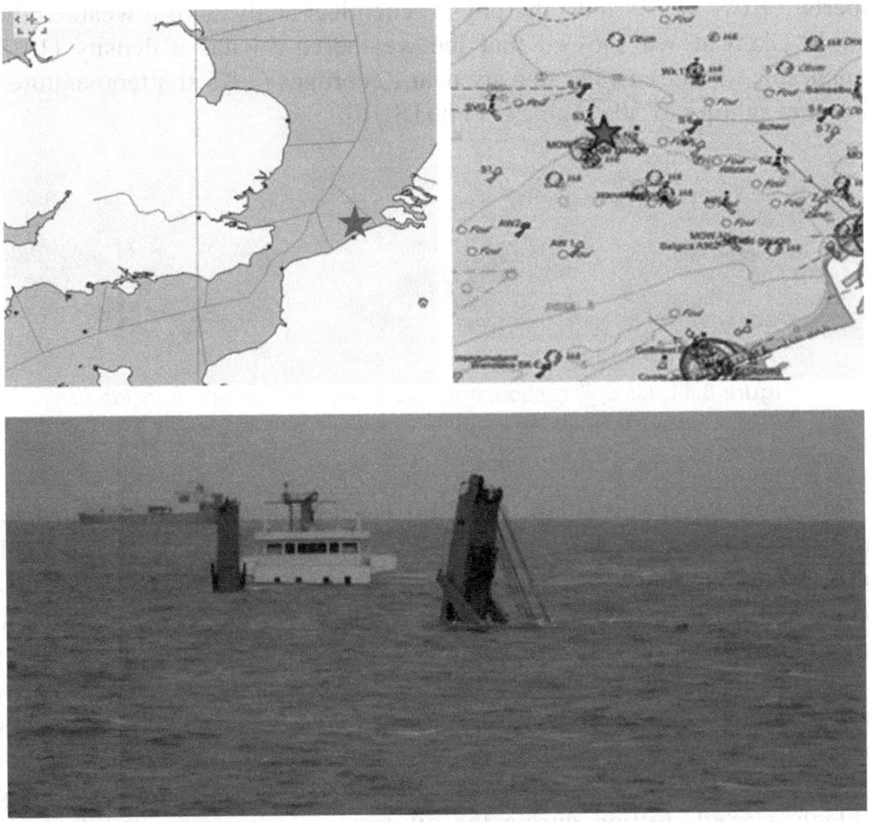

Figure 8.10. *Location of the* Flinterstar *wreck*

There was an almost continuous release of fuel oil from the wreck of the *Flinterstar*, from the start of the incident on October 6 until the end of the fuel removal operations on November 2, 2015, requesting the activation of three oil recovery vessels with sweeping arms (Figure 8.11). It is worth noting that booms and skimmers activated on the first day of the pollution and deployed in a U-configuration were unable to contain the HFO, probably due to the strong tidal currents in the area and the strange behavior of the apparently quite fluid HFO, which got mixed with gasoil. Furthermore, it was observed that the spilt HFO sometimes had a tendency to disappear from the sea surface and to drift sub-surface: thicker oil films consisting of true oil color were mainly observed nearby the wreck site, although the observed oiled (sheen) area often amounted to several square kilometers. Moreover, in calmer wind conditions, the previously submerged HFO sometimes tended to reappear at the surface. This "yoyo" effect puzzled the experts for two weeks until the physicochemical analysis of a weathered oil sample taken at sea showed that the weathered oil has a density (1.011) comparable to the seawater density near Zeebruges (1.02 at a temperature of 14°C and salinity of 30) (Table 8.1) [OTS 16].

Figure 8.11. *Oil spill response at sea during the* Flinterstar *incident: (a)* Arca*; (b)* Interballast I*; (c)* Hebocat 7

Based on aerial surveillance and oil recovery data, it is estimated that ca. 200 m^3 of fuel oil (mixtures of HFO and gasoil) was released in the marine environment over that 1-month period and ca. 50 m^3 was mechanically recovered. More precisely, for the first week following the initial collision, an oil spill in a continuous order of magnitude of 50–100 m^3 was observed at the sea surface (Figure 8.12); although fresh oil releases were continuously observed from the *Flinterstar* wreck, the volumes observed in the second week decreased drastically (with estimated volumes in the order of 1–10 m^3). At the end of the second week however, cargo suddenly started shifting during the oil removal operations, resulting in a pierced fuel tank and new significant releases of oil from the *Flinterstar* wreck, leading to observed oil volumes in the order of 50 m^3 or more. Thereafter, only minor oil releases (order of 1 m^3 or less) were observed until the end of pumping operations on November 2, 2015.

	HFO tank 1	HFO tank 3	HFO tank 4	*Arca* sample
Physical properties				
Viscosity at 15°C (mPa.s)	1982	1,496	2,150	78,000
Density at 15°C	0.991	0.955	0.993	1.011
Water content (%)	0%	0%	0%	61%
Chemical composition				
Content (% wt.)				
- SARA analysis:				
Saturates	34.7	48.5	32.8	37.3
Aromatics	33.7	28.2	32.3	30.2
Resins	19.8	19.5	24.9	19.9
nC_5-asphaltenes	11.7	3.8	10.0	12.6
- Asphaltenes and waxes:				
nC_7-asphaltenes	8.3	1.7	9.3	11.9
Waxes	7.1	2.6	7.0	5.5

Table 8.1. *Physicochemical characterization of HFO reference samples (Ref.: CEDRE RPTs L15.25 and L16.08)*

8.5.2. *Monitoring and surveillance of the oil and its fate/behavior*

Given the gravity of the situation, which was not catastrophic but severe (total loss of a vessel; significant maritime safety and environmental impact risks; oil spill of >100 m^3), Belgium's national contingency plan for the North Sea – the so-called North Sea General Emergency and Intervention Plan – was immediately activated. The environmental monitoring and surveillance programs involved *in situ* sampling (oil and seabed), remote sensing observations and modeling.

For instance, in the period between October 6 and November 2, 2015, 40 consecutive flights, or 55 "on task" flight hours, were performed by the remote sensing aircraft for aerial monitoring of the oil pollution and aerial assistance of response units. Further airborne assets that were deployed for aerial monitoring of the spill and incident site consisted of a helicopter and UAVs. Four CleanSeaNet satellite images were received on a regular basis offering valuable snapshots of the overall extent of the accidental oil pollution and resulted in a total of nine probable oil spill detection alerts that were received through CleanSeaNet and were verified and confirmed as mineral oil in the field by the Belgian authorities.

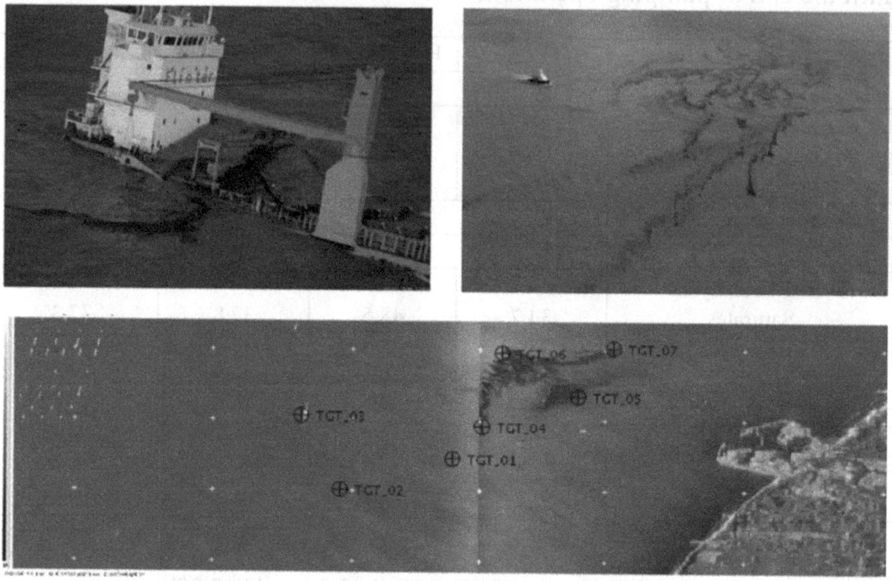

Figure 8.12. *Illustrations of* Flinterstar *wreck and the significant initial oil releases on October 6, 2015: SLAR image and photos taken from surveillance aircraft*

Oil samples were taken in the field throughout the incident period. Reference samples of the HFO were first obtained from the *Arca* (sample of recovered HFO taken onboard *Arca*); later on, during the oil removal operations, reference samples were also obtained from the different HFO fuel tanks. The fingerprinting analyses performed on these samples by the RBINS chemical lab in accordance with the CEN methodology could positively match the field samples taken at sea and on the beaches with the reference samples, demonstrating a clear causal link. The fingerprinting analysis performed on the reference samples also indicated that at least part of the HFO that was pumped out of the wreck a couple of weeks after the start of the incident had already weathered to a certain degree inside the wreck.

Reference samples were also sent to CEDRE for a detailed physicochemical characterization of the HFO present on board the *Flinterstar* to obtain a better understanding of the strange behavior of the heavy fuel when released into the sea and to improve released oil volume estimations.

Since the incident took place in very shallow and vulnerable coastal waters and the spilt oil posed an immediate risk for the nearby Belgian and Dutch shores, OSERIT was activated to evaluate and predict the drift and behavior of the spilt oil as well as possible. Based on factual data (known location of the wreck and information contained in the *Flinterstar*'s last bunker sheet), a series of 3D simulations have initially been performed assuming continuous release from a partially immerged ship wreck of fresh HFO-300 and MDO (marine diesel) and validated against the daily pollution observation reports from the various surveillance platforms (Figure 8.13).

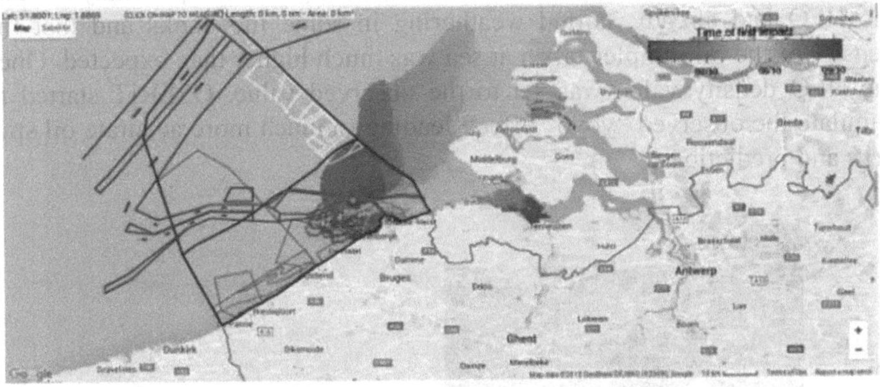

a)

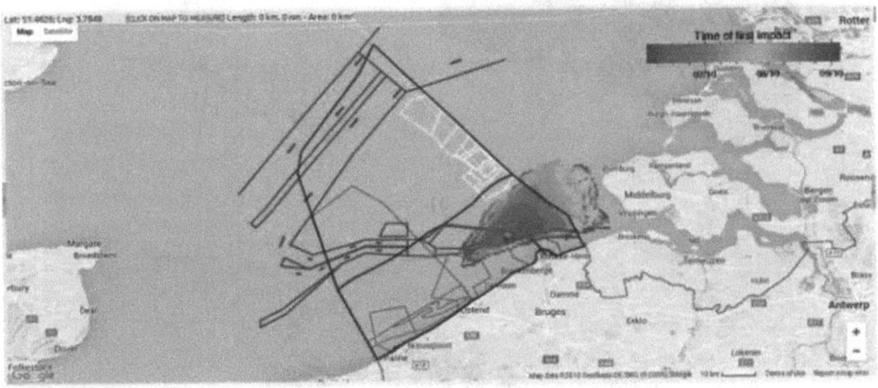

b)

Figure 8.13. *OSERIT 3-day oil drift model simulation for the period October 6 to October 9, 2015, of 75 m^3 of fresh HFO-300: (a) area impacted by surface slick(s); (b) area impacted by naturally dispersed oil. For a color version of the figure, see www.iste.co.uk/lefloch/remote.zip*

Although they provided useful information for the response planning, this first series of simulations were not able to simulate and explain the observed "yoyo" effect which lead to overestimation of the oil fraction remaining at the sea surface, an overestimation of the sea surface area impacted by the surface oil slicks and, in the case of the North-East wind, an overestimation of the beaching risk (i.e. false alert for the civil protection in charge of the beach cleaning operations). Although dozens of calibration simulations were performed to try to adjust the discharge rate at the breach, the natural dispersion rate and the oil distribution size droplets, the simulations were not totally satisfying until it was demonstrated, through oil sample analyses, that the HFO had already started weathering in some fuel tanks and that the density of the oil samples taken at sea was much higher than expected. Once the HFO density value was set to the observed value, OSERIT started to simulate the observed "yoyo" effect, leading to much more accurate oil spill drift and predictions.

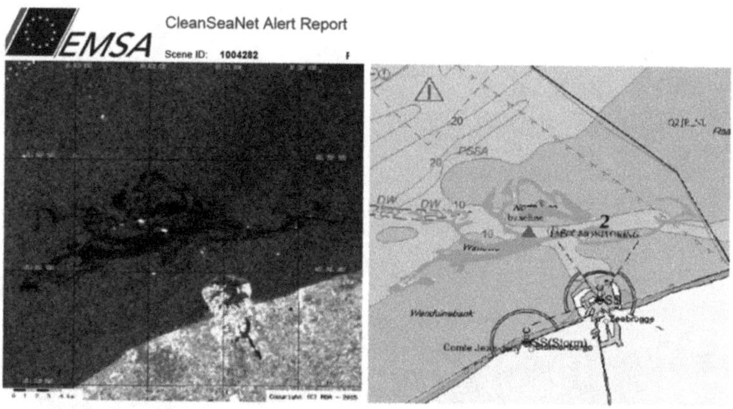

a)

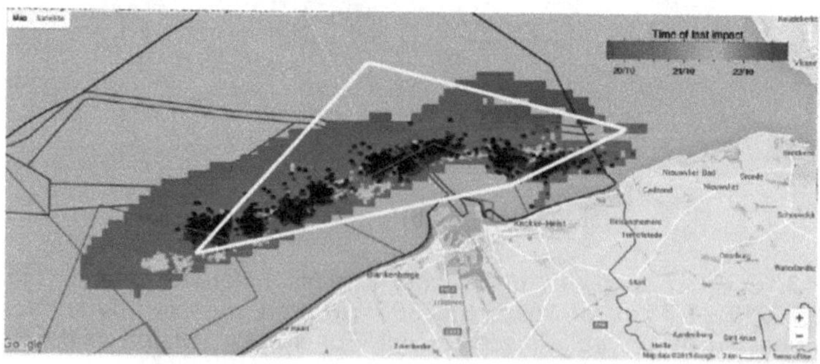

b)

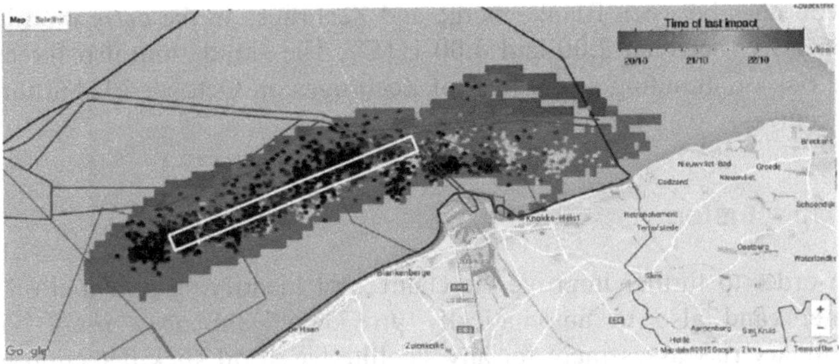

c)

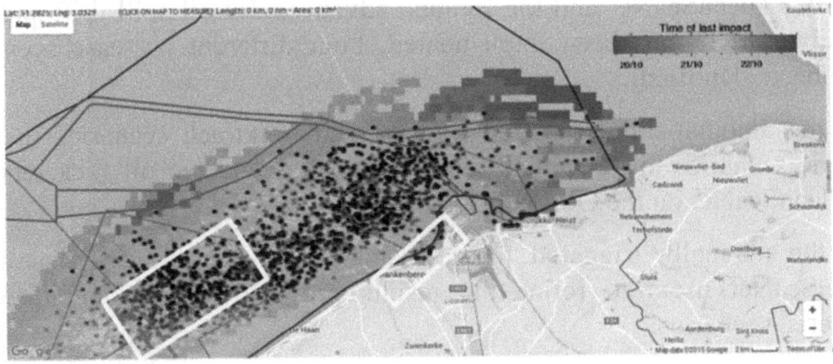

d)

Figure 8.14. *(a) Example of manually triggered simulation results of the system-to-system interface between CleanSeaNet and OSERIT for CleanSeaNet scene ID 1004282 (October 19, 17:30 UTC) at time steps for which aerial observations were available (yellow rectangle): (b) for flight 15179 (October 20, 07:00), (c) flight 15181 (October 20, 12:30 UTC) and (d) flight 15182 (October 21, 6:30 UTC). The black dots show the simulated position of the Lagrangian particles at the sea surface at the specified time; dark gray dots, one hour earlier, and light gray dots, two hours earlier. For a color version of the figure, see www.iste.co.uk/lefloch/remote.zip*

Although the system-to-system interface was not yet operational at that time, a preliminary version of the interface has been applied on the CleanSeaNet scene ID 1004282 (October 19, 17:30 UTC) (Figure 8.14 (a)). The simulation corresponds to a manually triggered simulation in which the oil density has been set to 1.011. The simulation results perfectly matched the aerial observations made during surveillance flights 15179 (October 20, 7:00 UTC), 157981 (October 20, 12:30 UTC) and 15182 (October 21, 6:30 UTC) and could partially explain the minor stranding of tar balls at the west of Zeebruges between Blankenberge and Zeebruges in the early morning of October 21 (between 2:00 and 4:00 UTC). The simulation also forecasted some risk of stranding at the East of Zeebruges on October 21, but this did not occur.

8.6. Conclusion

In order to further improve the chances of identifying potential oil spill polluters and also to automatically provide CleanSeaNet users with a preliminary risk assessment for any satellite-based oil slick detection in the North Sea and the English Channel, a system-to-system interface between EMSA's CleanSeaNet and the Belgian advanced 3D oil spill drift and fate model OSERIT has been implemented. Four different use-case scenarios have been identified:

– the automatically triggered forecast and backtrack scenarios that are generic scenarios to be applied to every satellite-detected oil slick reported by CleanSeaNet on the OSERIT domain;

– the manually triggered forecast and backtrack scenarios that allow CleanSeaNet users to refine the results provided by the automatically triggered scenarios.

CleanSeaNet users have had access to this new system from early 2017 onwards and the preliminary and test results have already demonstrated the usefulness of the system-to-system interface. During the *Flinterstar* incident in 2015, SAR images were successfully used, in combination with other aerial observations at the time of the incident, to validate the OSERIT model simulations and demonstrate their accuracy.

8.7. Acknowledgments

The implementation of the system-to-system interface between CleanSeaNet and OSERIT has been financed by the contract EMSA/NEG/69/2013. Sébastien Legrand warmly acknowledges Anne Marie Hayes, Gianluca Luraschi and Jose Matos from the European Maritime Safety Agency's Unit C – Operations as well as Nino Pace from Advanced Computer Systems for their active collaborations.

8.8. References

[BON 16] BONN AGREEMENT SECRETARIAT, Inventory of assessment tools (computer models), Bonn Agreement Counter-pollution Manual, Bonn Agreement Secretariat, London, 2016.

[DUL 13] DULIÈRE V., OVIDIO F., LEGRAND S.,"Development of an integrated software for forecasting the impacts of accidental oil pollution – OSERIT report, Belgian Science Policy Office, Brussels, 2013.

[FAY 71] FAY J.A., "Physical processes in the spread of oil on a water surface", *Proceedings of the Joint Conference on Prevention and Control of Oil Spills*, Washington, D.C., USA, June 15–17, 1971.

[FIN 96] FINGAS M., "The evaporation of oil spills: variation with temperature and correlation with distillation data", *Proceedings of the 9th Arctic Marine Oil Spill Program*, Calgary, Canada, pp. 29–72, 1996.

[FIN 97] FINGAS M., "The evaporation of oil spill: prediction of equations using distillation data", *Proceedings of the 20th Arctic and Marine Oil Spill Program*, Vancouver, Canada, pp. 1–20, 1997.

[FIN 11] FINGAS M. (ed.), *Oil Spill Science and Technology: Prevention, Response and Clean-Up*, Elsevier and Gulf Professional Publishing, Boston, 2011.

[FRE 03] FRENCH MCCAY D., "Development and application of damage assessment modelling: example assessment for the North Cape oil spill", *Marine Pollution Bulletin*, vol. 47, nos 9–12, pp. 341–359, 2003.

[GAR 99] GARCÍA-MARTÍNEZ R., FLORES-TOVAR H., "Computer modelling of oil spill trajectories with a high accuracy method", *Spill Science and Technology Bulletin*, vol. 5, no. 5/6, pp. 323–330, 1999.

[JOK 99] JOKUTY P., WHITICAR S., WANG Z. *et al.*, Properties of crude oils and oil products, Report no. EE-165, Environment Canada, Ottawa, Canada, 1999.

[JON 97] JONES R.K., "A simplified pseudo-component oil evaporation model", *Proceedings of the 20th Arctic and Marine Oil Spill Program*, Vancouver, Canada, pp. 43–61, 1997.

[LAG 12] LAGRING R., DEGRAER S., DE MONTPELLIER G. *et al.*, "Twenty years of Belgian North Sea aerial surveillance: a quantitative analysis of results confirms effectiveness of international oil pollution legislation", *Marine Pollution Bulletin*, vol. 64, no. 3, pp. 644–652, 2012.

[LEH 84] LEHR W.J., FRAGA R.J., BELEN M.S. *et al.*, "A new technique to estimate initial spill size using a modified Fay-type spreading formula", *Marine Pollution Bulletin*, vol. 15, no. 9, pp. 326–329, 1984.

[MCA 87] MCAULIFFE C.D., "Organism exposure to volatile/soluble hydrocarbons from crude oil spills – a field and laboratory comparison", *Proceedings of the International Oil Spill Conference*, Baltimore, USA, April 6–9, 1987.

[OSE 20] OSERIT, Oil Spill Evaluation and Response Integrated Tool, Mumm and Belgian Federal Science Policy Office, available at: https://oserit.naturalsciences.be (accessed June 2020), 2020.

[OTS 16] OTSOPA, The Flinterstar incident. report no. 16/8/Info.1 (L), OTSOPA 16, Scheveningen, The Netherlands, May 24–26, 2016.

[SCO 05] SCORY S., "The use of mathematical models for estimating oil pollution damage at sea", in MAES F. (ed.), *Marine Resource Damage Assessment: Liability and Compensation for Environmental Damage*, Springer, Netherlands, 2005.

[TKA 02] TKALICH P., CHAN E.S., "Vertical mixing of oil droplets by breaking waves", *Marine Pollution Bulletin*, vol. 44, no. 11, pp. 1219–1229, 2002.

[WAN 08] WANG S.D., SHEN Y.M., GUO Y.K. *et al.*, "Three-dimensional numerical simulation for transport of oil spills in seas", *Ocean Engineering*, vol. 35, nos 5–6, pp. 503–510, 2008.

9

Optimizing the Use of Aerial Surveillance Assets in Oil Spill Response Operations

9.1. Introduction

Offshore oil spill response operations, and more generally any maritime operation requiring a high level of situational awareness, often rely on the use of aerial assets to continually monitor the scene. The near-constant presence of aircraft on scene presents both opportunities (including but not limited to the ability to continuously monitor the situation as it develops, update the on-scene coordinator (OSC) and evaluate the effectiveness of operations on an ongoing basis) and challenges. While there are costs associated with the use of aircraft (both pecuniary and human), the main challenge is of a logistical nature. The availability of aircraft for such operations, the supply train that they rely on and the relative priority of other concurrent operations also factor in the frequency of their overflights, thus potentially limiting the "refresh rate" available to the OSC. This "refresh rate" has obvious effects on their ability to provide a bird's eye view of operations that can, in turn, have severe consequences on the ability of an OSC to effectively manage oil spill response operations. The wholesale development in the availability, at a fraction of the cost previously associated with such systems, of small unmanned aircraft systems (SUAS) in tandem with the advent of cheaper, miniaturized high-definition visual and multi-spectral sensors renders the use of SUAS as a supplementary source of airborne surveillance a very attractive proposition indeed. In order to

Chapter written by Charles Henri THOUAILLE.

evaluate the benefits of SUAS in oil spill response operations at sea, CEPPOL has been running a campaign of tests since November 2015.

Extensive research efforts have been underway for a decade on the use of unmanned aircraft systems (UAS) in the oil spill response sector [ENG 99, FIN 97, LAU 12, LON 12]. This research, while focusing mostly on military grade UAS, supports the thesis that UAS use in oil spill response (OSR) operations provides "added value".

> "If a suspect oil leak is detected, the UAS can be [...] vectored to the suspect sight where it can loiter and video record the situation with the side-mounted color or IR camera, providing direct GPS position and visual information to the field control station [...]. A single trained operator of a two man team is all that is required to review and asses the real time information displayed." [ALL 08]

Our aim in undertaking this experimental campaign is not to compete with the pre-existing body of research on the topic but rather to build on what has already been widely acknowledged, namely that "UAS show promise as a tool in oil spill response observation and spotting [and] can provide command and control with additional real-time situational awareness" [PAR 14] and test whether these conclusions hold true when transposed to commercially available (off-the-shelf) consumer-oriented SUAS.

With the advent of cheap, mass-produced, consumer-oriented "drones", the availability of relatively easy-to-use aerial sensor systems has exploded. Their small size, ease of use and ability to carry light payload (such as high definition video cameras) for up to 25 minutes while streaming back high-quality video signals in real time indicate that with little training, most people can operate airborne sensors.

In a field where situational awareness is the only way to ensure the effectiveness of operations, having such sensor systems at our disposal seems a comparatively cheap way to complete airborne surveillance operations when other, more advanced aircraft (such as specialized helicopters or airplanes) are not available.

While on a purely theoretical basis, the characteristics of commercially available SUAS, as announced by their manufacturers, seem promising, an *in situ* experimentation protocol seems necessary in order to determine the fitness for use of off-the-shelf systems.

9.2. Assumptions and working hypotheses

Tables 9.1–9.3 outline the assumptions at the core of the working hypotheses used in our experimental protocol.

A1: Core assumption	**A1: Off-the-shelf SUAS offer added value in offshore and near-shore oil spill response operations.**
A1: Corollary assumption	**A1(i):** Flight time and operational radius are sufficient for close-range scouting and reconnaissance missions.
	A1(ii): Payload available will successfully be able to detect slicks and stream visual data in real time to the SUAS operator.
	A1(iii): As close-range, supplementary sensors, SUAS will enable better situational awareness for on-scene commanders..

Table 9.1. *A1 assumptions*

A2: Core assumption	**A2: Off-the-shelf SUAS are deployable at sea, with sufficient ease that the potential disruption they may cause on normal operations (cost) is far outweighed by the added situational awareness (benefit).**
A2: Corollary assumption	**A2(i):** SUAS operations are compatible with oil spill response operations as carried out onboard OSR vessels.
	A2(ii): Deployment and recovery are sufficiently easy to execute that pilots require minimal training (low investment/high yield) and the training time required to operate SUAS effectively is reasonably short.
	A2(iii): Training and associated costs are reasonably low.
	A2(iv): The extant legal framework allows the use of SUAS in operational scenarios compatible with their intended use.

Table 9.2. *A2 assumptions*

A3: Secondary assumption	**A3: Off-the-shelf sensors, compatible with SUAS use, provide a human operator the source material to reliably identify, qualify and quantify oil spills remotely.**
A3: Corollary assumption	**A3(i):** Visible spectrum cameras commercially available off the shelf are fit for this purpose.
	A3(ii): Multi-spectral sensors commercially available off the shelf are fit for this purpose.
	A2(iii): Thermal infrared sensors commercially available off the shelf are fit for this purpose.
Core hypothesis	Off-the-shelf SUAS are able to successfully detect slicks at sea as close-range airborne surveillance systems, while their ease of use ensures that the cost–benefit analysis remains firmly on the positive side of the balance.
Secondary hypothesis	Off-the-shelf miniaturized imaging sensors (visible and invisible spectrums) can be used in conjunction with SUAS to provide OSCs with better situational awareness.

Table 9.3. *A3 assumptions and working hypotheses*

The experimental protocol currently underway at CEPPOL is designed to test the working hypotheses by testing the validity of each core assumption through the step-by-step validation of each one of their corollaries.

9.3. Experimental protocol: testing the primary hypothesis

Prior to the decision to pursue a campaign of *in situ* tests with a view to determining the usefulness of SUAS in oil spill response operations, it was necessary to start with a literature review. Research sources cited in the bibliography were reviewed in parallel with information available from recreational "drone" manufacturers and retailers.

9.3.1. *Technical specifications*

On the basis of the technical review, CEPPOL defined the following technical specifications that appeared best suited for compatibility with the current operational framework and standard operating procedures.

Flight parameters	Image	Others
P1: Assisted flight modes (automatic management of stability, attitude, altitude, speed)	P1: Ability to stream back high-quality video to operator for slick detection	P3: Waterproof P2: Flight parameters and telemetry available remotely in real time
P1: Automatic stationary flight mode (GPS-enabled)	P1: HD video recording	P1: Does not require a license in order to operate
P1: Autonomy (18+ minutes)	P1: HD photography	P1: Available off the shelf for under 5,000 €
P1: Operational altitude: >150 m	P1: Stabilized image (gimbal or software stabilization)	P2: Able to embark additional payload
P1: Max distance from operator >1,000 m	P2: Can be customized (upgrade camera/change sensors)	P1: Ground station, drone and spare parts fit in a backpack
P1: Wind resistance: >20 kts	P2: Images can be geotagged	P2: Ability to track SUAS position (GPS coordinates) in real time
P2: Operating temperature range: −10°C to +40°C		
P2: Ability to return to operator GPS coordinates		

Table 9.4. *Technical specifications*

9.3.2. *Operational requirements*

For SUAS to be of use, they must be compatible with the existing standard operation procedures (SOP) of the organization that will be using them. If not, the adaptations of said SOPs must be minimal. For instance, in order for CEPPOL to find an *a priori* interest in SUAS as a supplementary airborne surveillance capacity, the following minimum operational requirements have to be met:

– ability to operate SUAS without embarking additional personnel when conducting spill response operations;

– ability to deploy and recover the SUAS from a cluttered deck onboard a seagoing vessel;

– ability to operate SUAS in weather conditions compatible with spill recovery operations (up to sea state 3–4/winds up to 30 kts max);

– deployment and recovery of SUAS do not interfere with spill cleanup operations;

– SUAS operations do not constitute an additional hazard for personnel involved in ship conduct or spill cleanup operations.

These criteria echo some of the difficulties outlined in [PAR 14], where it was found that operating SUAS from sea going vessels is possible, but the nature of OSR vessels and the equipment they carry present both opportunities (deck space) and pitfalls (large amount of equipment occupying the space). With a view to facilitating operations, we decided to start our investigations with vertical takeoff and landing (VTOL)-capable SUAS, leading us to begin our investigations with quadcopters.

9.3.3. *Choice of SUAS*

The technical specifications and operational requirements listed in the previous paragraphs served to narrow the field of commercially available systems. A number of contenders were available at the time of purchase, but none that would fit all criteria listed above were then available off the shelf. The technical specifications were further prioritized and CEPPOL acquired a 3DR Solo [3DR 15] in order to proceed with the *in situ* experimentation phase. This particular model met all P1 technical specifications except the minimum range of 1,000 m with live HD video streamed back to the pilot. Given the expected conditions of use (close-range slick repositioning, assessment of the effectiveness of spill response operations, increased situational awareness to aid OSC in coordinating vessels, etc.), a distance of 800 m from the operator while maintaining live video stream was judged sufficient. Appendix 2 provides an overview of the technical specifications deemed: essential (P1), recommended (P2) and desirable (P3) and matches them against Solo's specifications.

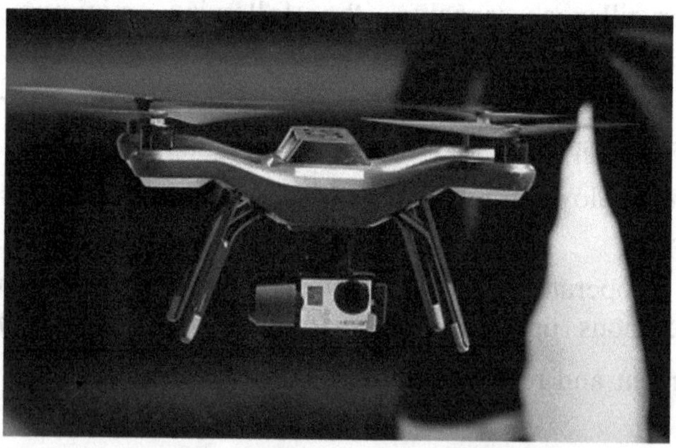

Figure 9.1. *CEPPOL SUAS (3DR Solo and GoPro 3+Black)*

9.3.4. *Systematic testing of assumptions*

The core hypothesis formulated during the review process was as follows:

> Off-the-shelf SUAS are able to successfully detect slicks at sea as close-range airborne surveillance systems while their ease of use ensures that the cost–benefit analysis remains firmly on the positive side of the balance.

In order to determine its validity both core assumptions and their corollaries were tested in turn. For the most part, these tests were carried out during oil spill training deployments (half-day to a day at sea where various types of equipment are deployed in order to maintain personnel and equipment readiness levels). However, three large-scale national oil spill response exercises provided an excellent opportunity to test some of the assumptions in a realistic scenario where SUAS operations had to "fit in" the operational response or be abandoned. These exercises showed, as a preliminary conclusion, that the use of off-the-shelf SUAS was compatible with the current SOPs and the concept of operations without major evolutions.

To date, all tests performed in determining the fitness for purpose of SUAS as a supplementary airborne sensor to be used in oil spill response operations have been carried out using simulants (rice husks and fluorescein) to model oil slicks. In the absence of actual spills during the various tests, the airborne sensors that equip CEPPOL's SUAS have not been tested with real oil. Nevertheless, the preliminary conclusions that will be drawn from these tests can then be assessed against a wide body of data available following oil spill response operations worldwide.

9.3.4.1. *Evaluating core assumption 1*

A1: Core assumption	**A1: Off-the-shelf SUAS offer added value in offshore and near-shore oil spill response operations.**
A1: Corollary assumption	**A1(i):** Flight time and operational radius are sufficient for close-range scouting and reconnaissance missions.
	A1(ii): Payload available is able to successfully detect slicks and stream visual data in real time to the SUAS operator.
	A1(iii): As close-range, supplementary sensors, SUAS will enable better situational awareness for on-scene commanders.

Table 9.5. *A1 evaluation*

Since acquisition, CEPPOL's SUAS has been deployed at sea on six occasions to test the validity of the assumptions that support our core hypothesis. Three of these deployments were short training exercises and three of these were large-scale national exercises.

Assumption A1 rests on the corollaries listed in the table above. Flight time and operational radius have proved sufficient to successfully survey the area of operations at close range around the OSC's oil spill response vessel (OSRV) and other units deployed to combat simulated oil spills.

The payload available at the time enabled the OSC to verify that the assets they had tasked were indeed deployed in accordance with the operation plan and that they were deployed as effectively as possible. One of the reasons airborne surveillance assets are necessary when dealing with oil spills is the difficulty for crews on a vessel's bridge to accurately see a slick. When close to the sea level, a slick's exact location and size is difficult to identify with any degree of precision. SUAS allow the on-scene coordinator a bird's eye view, thus enabling them to see whether the assets at their disposal are in or next to the slick they are tasked with recovering.

Further testing with real oil and upcoming tests with different payloads (including multi-spectral and thermal infrared sensors) are expected to further support this preliminary finding. A wide body of evidence exists to back up the part of our core hypothesis that relies on off-the-shelf optical sensors being able to distinguish oil spills from the air. Technical findings that stem from similar research conducted with different types of UAS also support this hypothesis. In their 2014 paper, Parscal and Ziska provide clear evidence that visual spectrum sensors carried onboard "regular" UAS can and do detect oil slicks [PAR 14]. While the UAS they used in their research and CEPPOL's 3DR Solo differ in size and form, their payload and flight parameters are sufficiently close that we posit that their UAS' ability to detect oil provides sufficient evidence to support corollary A1(ii).

Despite not having been able to test our SUAS against real oil slicks, we consider that the existing research proves this assumption to be sound.

Corollaries A1(i) through A1(iii) are thus deemed correct. Furthermore, during the maritime prefecture for the English Channel and North Sea's annual oil spill response exercise, a dense fog covered the entire area of offshore operations. No manned aircraft was permitted to operate due to security concerns. CEPPOL's SUAS suffered no such constraints; it was successfully deployed and recovered after two consecutive missions, providing the OSC with valuable information (including nearby slick positioning), further proving that the use of SUAS in oil spill response operations could provide added value.

Figure 9.2. *Visual detection of rice husk from 100 m altitude (SUAS: 3DR Solo & GoPro 3+Black)*

9.3.4.2. *Evaluating core assumption 2*

A2: Core assumption	**A2: Off-the-shelf SUAS are deployable at sea, with sufficient ease that the potential disruption they may cause on normal operations (cost) is far outweighed by the added situational awareness (benefit).**
A2: Corollary assumption	A2(i): SUAS operations are compatible with oil spill response operations as carried out onboard OSR vessels.
	A2(ii): Deployment and recovery are sufficiently easy to execute that pilots require minimal training (low investment/high yield) and the training time required in order to operate SUAS effectively is reasonably short.
	A2(iii): Training and associated costs are reasonably low.
	A2(iv): The extant legal framework allows the use of SUAS in operational scenarios compatible with their intended use.

Table 9.6. *A2 evaluation*

Core assumption 2 has been less extensively tested. While corollaries (i)–(iii) have been proven correct in the course of our various tests, the constant evolution of the French legal framework dealing with the use of SUAS proves more elusive.

As evidenced by the repeated use of the SUAS in national exercises, SUAS operations are compatible with spill response operations carried out on specialized vessels. The space available for deployment and recovery is sufficient; these operations do not disrupt the normal operations that take place on deck or on the bridge, and while rough weather conditions can make recovery of the SUAS challenging at times, most responders with regular practice should be able to successfully deploy and recover SUAS. Thus far, only one pilot has carried out live deployments at sea, but all CEPPOL technical staff have had opportunities to practice flying in challenging conditions ashore.

Training an operator to operational standards (ability to deploy and recover the SUAS, ability to perform basic search patterns) can be achieved in a matter of weeks. Anyone can successfully perform basic search patterns at close range. However, recent evolutions in the French legal framework make official training practically unavoidable for personnel likely to operate SUAS at sea in oil spill response operations.

At this stage, SUAS deployment and recovery operations have only been conducted up to winds of 20–25 kts. The maximum wind conditions have not been tested thus far. Unpredictable variations in wind force can have a marked impact on the trajectory of SUAS during deployment and recovery where the environment of an OSRV rear platform (largely occupied by spill response equipment) increases the chance of a crash or loss of the SUAS. Operator confidence levels and regularity of training must therefore be considered before deploying in less clement conditions. While corollary A2(ii) is considered proven correct, deployment and recovery operations, at sea, from a vessel equipped with the necessary equipment for oil spill response operations are highly technical and necessitate some practice. The relative difficulty of those operations must not be underestimated.

The financial cost of training is a known quantity. If an operator needs to obtain a professional license, then the cost associated with training varies, but the mean price of a week-long one-on-one course represents an expenditure of approximately 2,500 €. Given the additional freedom afforded to formally trained and recognized operators under French law, such a course is not deemed to be overpriced. Corollary A2(iv) is thus deemed provisionally proven.

9.3.4.3. *Preliminary findings*

Tests carried out to this day support the validity of the core assumptions:

> Off-the-shelf SUAS are able to successfully detect slicks at sea as close-range airborne surveillance systems, while their ease of use ensures that the cost–benefit analysis remains firmly on the positive side of the balance.

The core hypothesis recalled above therefore appears to be supported by the evidence collected in the course of CEPPOL's experimentation. Further experimentation will be necessary to determine the exact place of SUAS in oil spill response operations and what, if any, limitations appear when dealing with real oil.

9.4. Experimental protocol: underlying assumptions and testing of secondary hypothesis

The secondary hypothesis (recalled below) rests on much the same core assumptions as the primary hypothesis. In fact, it takes as a given the validity of the core hypothesis. Indeed, if our core hypothesis had proven unfounded, then the secondary hypothesis would be null and void. Only three additional assumptions are made in reaching our secondary hypothesis. Those are recalled below as well.

A3: Secondary assumption	**A3: Off-the-shelf sensors, compatible with SUAS use, provide a human operator with the source material to reliably identify, qualify and quantify oil spills remotely.**
A3: Corollary assumption	**A3(i):** Visible spectrum cameras commercially available off the shelf are fit for this purpose.
	A3(ii): Multi-spectral sensors commercially available off the shelf are fit for this purpose.
	A2(iii): Thermal infrared sensors commercially available off the shelf are fit for this purpose.
Secondary hypothesis	Off-the-shelf miniaturized imaging sensors (visible and invisible spectrums) can be used in conjunction with SUAS to provide OSCs better situational awareness.

Table 9.7. *A3 evaluation*

Taking as a given that SUAS can successfully be used to detect slicks at close range, once their compatibility with oil spill response operations has been established, is not a farfetched assumption.

This assumption is supported by the findings of multiple studies, for instance [ALL 08, PAR 14, DEM 15]. In order to further substantiate this claim, two types of simulants were used. In July 2016, a field test involving the use of both rice husks and fluorescein was performed at sea. The SUAS (3DR Solo & GoPro 3+Black) successfully detected both simulants.

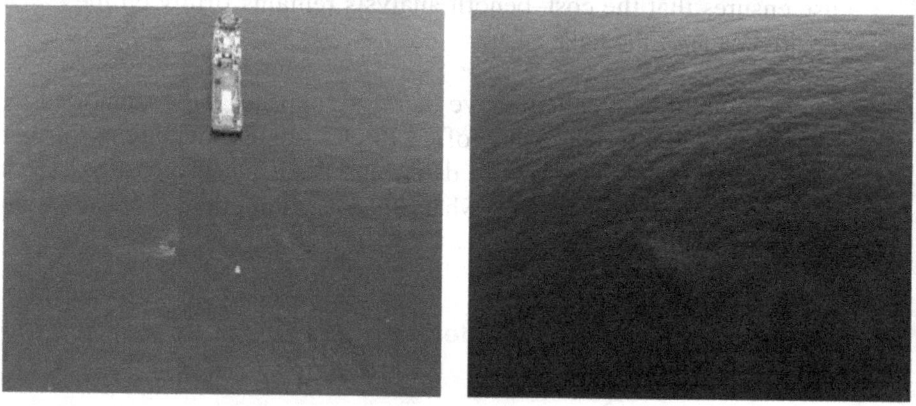

Figure 9.3. *Rice husks (left) and fluorescein (right) detection field tests (alt. 100 m), July 2016*

Organizations such as the Bonn Agreement rely on visual spectrum observation to detect, characterize and evaluate (quantify) oil spills [BON 08]. It is worth noting that the Bonn Agreement Appearance Code (BOAC) is used internationally for classifying pollutions and quantifying the oil present in a slick. Other leading organizations such as CEDRE and Oil Spill Response Limited also rely on the Bonn Agreement appearance code for this purpose [CED 04, OIL 11].

Given that qualification and quantification of pollutions on the basis of the BOAC is conducted both *in situ* by observers and *ex situ* on the basis of recorded images and that UAS bearing similar equipment (visible spectrum camera sensors) have been successfully proven to be well suited for oil spill detection, we consider that our field trials, despite the limitation of having been carried out without real oil, support corollary assumption A1(i).

A partnership with French drone maker Parrot will allow further testing of assumption A3 in the multi-spectral band. A multi-spectral sensor, initially developed for agricultural use, has been added to CEPPOL's SUAS, and initial tests of the device seem promising; however, at this stage, the first flights have only served to validate the viability (flight worthiness, maneuverability, etc.) of this addition to the SUAS.

The setup that we have devised will allow both regular color (visible) video to be streamed back in real time to the operator and access to high-resolution multi-spectral stills to a second operator. One factor that will limit the usefulness of this device is its inability to stream video in real time. Nonetheless, its ability to take pictures at a given rate (at a time interval defined by the operator) and to stream these back to a computer will mitigate this shortcoming. This has been tested in the course of our experimental protocol.

Figure 9.4. *CEPPOL SUAS experimental visible and multi-spectral bands*

However, the fitness for purpose of the Sequoia sensor [PAR 16] obtained in partnership with Parrot has not been tested. Based on the characteristics of the product as announced by the manufacturer, Sequoia operates in the following bands: full RGB spectrum, 550 nm wavelength/40 nm bandwidth (green), 660 nm/40 nm (red), 735 nm/10 nm (red edge) and 790 nm/40 nm (near infrared). The next field trials will be conducted in the last two bands (red edge and near infrared). However, these will need to be conducted in a controlled environment with real oil in order to be conclusive. The literature available on the subject of oil spill detection and classification in these bands [ADA 09] suggests that our future tests with Parrot's agricultural multi-spectral sensor should be successful. While further testing of corollary A3(ii) is necessary, we expect it to be successfully proven.

Our last corollary has not been tested yet. Thermal infrared imaging is known to be a highly effective tool in detecting oil spills and is widely used by coastguards to detect and monitor spills offshore [NAN 08]. As such, commercially available sensors such as the small and lightweight FLIR Vue (thermal infrared camera) that can be adapted onto SUAS for a minimal cost (preliminary assessments indicate that adding thermal IR capabilities to CEPPOL's SUAS would cost approximately 2,700 €) are expected to perform well and successfully prove corollary assumption A3(iii), thus providing the necessary evidence to prove our secondary hypothesis that off-the-shelf miniaturized imaging sensors (visible and invisible spectrums) can be used in conjunction with SUAS to provide OSCs with better situational awareness.

9.5. The case for using SUAS as a force multiplier in spill response coordination

The findings outlined throughout this chapter support the hypothesis that SUAS are not only compatible with oil spill response operations but also bring added value at a comparatively low price (under 5,000 €). Given the relative scarcity of specialized aircraft and therefore their availability for an OSC, the ability to supplement the airborne surveillance data made available by those aircraft with SUAS at immediate command appears highly desirable.

The view that to a large extent, oil spill response operations depend on the ability to identify and describe the characteristics of an oil spill and ultimately direct (and coordinate) the necessary resources for clean-up, and that this ability depends, in turn, on the presence and availability of trained personnel in manned aircraft, is shared by many in the oil spill response field [PAR 14].

The ability afforded by SUAS to frequently update the locations of oil patches, track their displacement over time (and confront it with drift models) and review, in real time, the effectiveness of spill response operations as posited by Allen and Walsh [ALL 08] is supported by our findings.

SUAS cannot replace regular surveillance assets as such. Their reach and endurance (among other things) do not permit long-term observation of the area of operations. Nevertheless, as a supplementary asset, capable of accelerating the rate at which information is refreshed and made available to OSCs, they seem to have a choice place in oil spill response operations.

As a tool for close-range situational awareness augmentation, SUAS do seem to constitute a force multiplier. By accelerating the refresh rate, they bring increased situational awareness to the on-scene coordinator, who can review situations for which this was previously impossible, assess the efficacy and effectiveness of their operations and thus better coordinate the assets made available to them.

The force multiplier effect is contingent on the good integration of SUAS operations in the wider OSR operations. Sea-to-air and air-to-air coordination is paramount so as to not run the risk of mid-air collision and optimize the use of resources. SUAS could therefore be made available to on-scene coordinators as an additional tool but only be deployed in the absence of other airborne surveillance assets. The extensive testing conducted during large-scale exercises also supports this claim.

9.6. Appendix 1

Abridged technical data sheet for CEPPOL SUAS.

MARINE NATIONALE

CEPPOL

CENTRE D'EXPERTISES PRATIQUES DE LUTTE ANTIPOLLUTION

Type of equipment : SUAS	Version: 01 24/11/2015	N°: 1
Designation: SOLO-CEPPOL-01	Manufacturer : 3D Robotics	

Description		
SUAS to be tested for deployment from oil spill response vessel in typical conditions (up to sea state 3-4 / wind < 30kts). **Additional equipment necessary** : GoPro Hero 3+Black (or compatible), tablet (IOS or Android), additional battery, ground station, battery charger, memory cards.		
Operational limits		
Maximum wind : 30kts Rain.	Technical data : • 20 minutes autonomy • Max dist. from pilot : 800m • Max altitude (legal) : 150m	Technical data (continued) : • Max speed : 40kts • Resolution : 12MP • Video 1080p at 60fps / 2.7k at 30fps
Characteristics		
• Live HD video streamed to pilot ground station / remote control. • Flight data & parameters (technical and geographic) logged and stored after each flight.		

Transport	Weight	Dimensions	Misc.
Backpack	1.8 kg	Length / Width: 46cm Height : 25cm	All necessary equipment in transport bag.

9.7. Appendix 2

Comparison of initial technical specifications identified with capabilities of 3DR Solo.

Flight parameters	Image	Others
P1: Assisted flight modes (automatic management of stability, attitude, altitude, speed)	P1: Ability to stream back high-quality video to operator for slick detection	P3: Waterproof **Not waterproof. Tested successfully in dense fog conditions**
P1: Automatic stationary flight mode (GPS-enabled)	P1: HD video recording	P2: Flight parameters and telemetry available remotely in real time ***Via a secondary interface only***
P1: Autonomy (18+ minutes)	P1: HD photography	
P1: Operational altitude: >150 m	P1: Stabilized image (gimbal or software stabilization)	P1: Does not require a license in order to operate
P1: Max. distance from operator >1,000 m **Actual distance announced and tested up to 800 m**	P2: Can be customized (upgrade camera/change sensors)	P1: Available off the shelf for under 5K €
P1: Wind resistance: >20 kts	P2: Images can be geotagged	P2: Able to embark additional payload
P2: Operating temperature range: −10°C to +40°C		P1: Ground station, drone and spare parts fit in a backpack
P2: Ability to return to operator GPS coordinates Only capable of automatic return to takeoff location		P2: Ability to track SUAS position (GPS coordinates) in real time ***Via a secondary interface only***

– white cells: specifications coherent with the initial specifications list;

– light gray cells: specifications approaching initial requirements;

– dark gray cells: initial specification not met but partly mitigated.

9.8. References

[3DR 15] 3D ROBOTICS, Solo User Manual, 3D Robotics, Berkeley, USA, 2015.

[ADA 09] ADAMO M., DE CAROLIS G., DE PASQUALE V. *et al.*, "Detection and tracking of oil slicks on sun glittered visible and near infrared satellite imagery", *International Journal of Remote Sensing*, vol. 30, no. 24, pp. 6403–6427, 2009.

[ALL 08] ALLEN J., WALSH B., "Enhanced oil spill surveillance, detection and monitoring through the applied technology of unmanned air systems", *Proceedings of the International Oil Spill Conference*, Savannah, USA, May 4–8, 2008.

[BON 08] BONN AGREEMENT SECRETARIAT, The Bonn Agreement Oil Appearance Code, Bonn Agreement Aerial Operations Handbook, Bonn Agreement Secretariat, London, 2008.

[CED 04] CEDRE, L'observation aérienne des pollutions pétrolières en mer, Operational guide, CEDRE Accidental Water Pollution, Brest, 2004.

[DEM 15] DE MICCO E.D., NEDWED T., PALANDRO D., "Small unmanned aerial systems to determine slick thickness", *Interspill Conference*, Amsterdam, 2015.

[ENG 99] ENGELHARDT F.R., "Remote sensing for oil spill detection and Response", *Pure Applied Chemistry*, vol. 71, no. 1, pp. 103–111, 1999.

[FIN 97] FINGAS M., BROWN C.E., "Review of oil spill remote sensing technologies", *Spill Science & Technology Bulletin*, vol. 4, no. 4, pp. 199–208, 1997.

[LAU 12] LAURENT M., "Fixed detection devices for oil slick", *CEDRE Information Days, Spill and Illegal Discharge Detection*, Brest, France, March 20, 2012.

[LON 12] LONG M.D., Remote sensing for offshore marine oil spill emergency management, security and pollution control, Presentation, Aerial Surveillance Systems Inc., Scottsdale, USA and Panama City, Panama, 2012.

[NAN 08] NAND JHA M., LEVY J., GAO Y., "Advances in remote sensing for oil spill disaster management: state-of-the-art sensors technology for oil spill surveillance", *Sensors*, vol. 8, no. 1, pp. 236–255, 2008.

[OIL 11] OIL SPILL RESPONSE, *Aerial Surveillance Field Guide*, 2nd ed., OSRL, Southampton, United Kingdom, 2011.

[PAR 14] PARSCAL B., ZISKA M., WILLIAMS J., "A field evaluation of unmanned aircraft systems for oil spill response", *Proceedings of the International Oil Spill Conference*, Savannah, USA, May 5–8, 2014.

[PAR 16] PARROT DRONES, Parrot Sequoia user guide, Parrot, 2016.

PART 4

Remote Sensing for Exploration

10

Potential of Imaging UAVs for Coastal Monitoring

10.1. Introduction

Metropolitan France has nearly 5,500 km of coastline. This coastal area concentrates many issues including environmental (great variety of geological features, flora, fauna, diversity of habitats, etc.) and societal stakes (housing, transport, tourism, fishing, industrial, energy, etc.). However, this coastal area has a strong and varied sediment dynamics, which is characterized by changes in the coastline, especially under the action of complex hydrodynamic and morphodynamic mechanisms. The National Observation Service DYNALIT (DYNAmics of LITtoral and coastline) was set up by the CNRS (Centre National de Recherche Scientifique) to further understand the dynamics of coastal areas. This service aims to collect and bring coherence to metrological data in the long term. With this aim in mind, 29 study sites were chosen (Figure 10.1). These sites are spread over all the French coastlines (metropolitan and ultramarine) and are representative of the different coastal contexts (sandy coasts, cliffs, estuaries). The objective is to understand the processes involved in the coastal dynamics. The parameters measured in this perspective are the topography, the bathymetry and the sea states.

Chapter written by Marion JAUD, Christophe DELACOURT, Nicolas LE DANTEC, Jérôme AMMANN, Philippe GRANDJEAN, Pascal ALLEMAND and Lucie COCQUEMPOT.

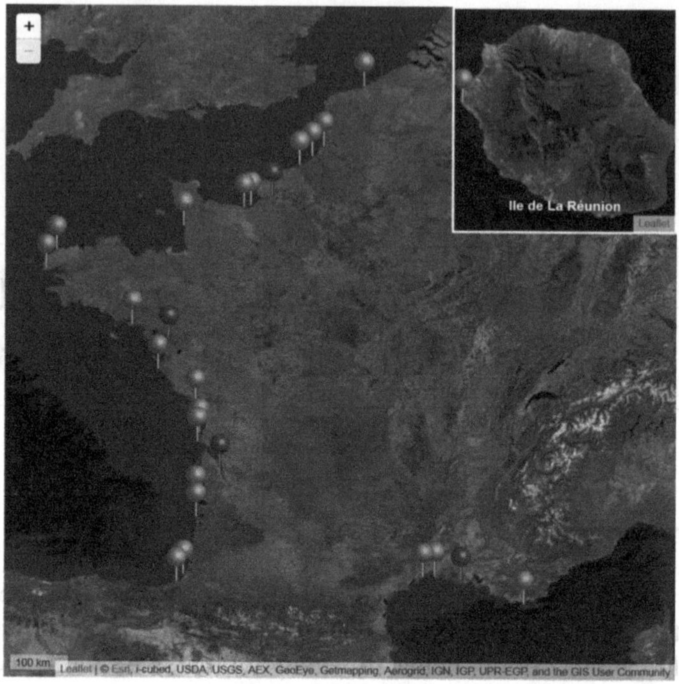

Figure 10.1. *Localization of DYNALIT study sites*

10.2. Constraints on the survey

The diachronic measurement of these parameters thus requires the development of specific tools and methods to meet specifications, particularly requiring:

– repeated acquisitions, with a frequency ranging from several days to several decades;

– flexibility in order to make surveys at key dates (tides, post-storm surveys, etc.) and to be able to adapt to weather constraints;

– limited acquisition time to adapt to tidal phenomena;

– ability to perform surveys on sites where access may be difficult;

– preservation of sites and respect for public space;

– equipment resisting coastal conditions (salt air, wind, spray, etc.).

For many years, the measurement of these parameters was performed by onboard sensors from ships, aircrafts or satellites, as well as by *in situ* measurements through DGPS cross-shore profiles.

Nevertheless, the development of new miniaturized image sensors (digital cameras, thermal sensors, hyperspectral, Lidar, etc.) and positioning systems associated with the development of new platforms such as unmanned aerial vehicles (UAVs) have led to the emergence of new coastal monitoring strategies. Indeed, these combined techniques are offering a synoptic vision, along with better spatial resolution and accuracy at lower costs.

Technological improvements of platforms and miniaturization of imaging and positioning sensors can meet these challenges.

10.3. Examples of UAV platforms

The Domaines Océaniques laboratory has been performing drone surveys of the coastline since 2006, by adapting the DRELIO system (Drone Helicopter Environment) to fit with the study issues and innovations offered by the market (Figure 10.2) [KLE 15, DEL 09].

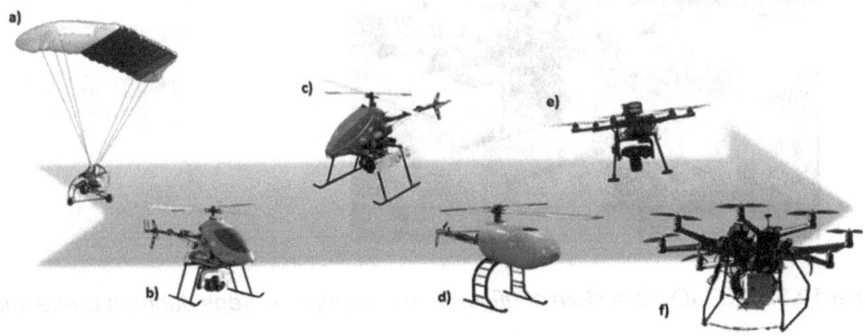

Figure 10.2. *Various models of drones implemented by the Domaines Océaniques laboratory since 2006: (a) Pixy wing; (b), (c), (d) thermal helicopters; (e) hexacopter; (f) octocopter*

The first drones used for coastal monitoring were flying Pixy wings developed by IRD (Research Institute for Development), requiring large areas of takeoff and little wind. Next, thermal engine helicopters were used with significant payload capacity, but which were difficult to pilot. Subsequently, these helicopters were equipped with autopilot. However, the revolution in the sphere of drones was the arrival of electric multicopter UAVs, which are more stable in flight, more reliable and easier to use. However, the carrying capacity of these UAVs is limited (a few hundred grams to several kilograms depending on the model), hence the miniaturization of sensors.

For photogrammetry, DRELIO 10, the model currently used by Domaines Océaniques, is an electric hexacopter UAV with a diameter of 0.8 m (Figure 10.3). This UAV is equipped with a collapsible frame that allows easy transportation. It weighs less than 4 kg and can handle a payload of 1.6 kg. The flight autonomy of DRELIO 10 is about 20 min. It is equipped with a reflex camera (Nikon D700) with a focal length of 35 mm, taking one 12 Mpix photo in intervalometer mode every two seconds. The camera is mounted on a tilting gyro-stabilized platform. The flight control is run by the DJI® software iOSD. Although DRELIO 10 is able to perform semi-autonomous flight, takeoff and landing, ground station software is used to control the UAV during the flight.

Figure 10.3. *DRELIO 10 system configuration and primary steps during a drone survey*

10.4. Survey protocol

In practice, targets should be installed over the studied area. These targets, whose position is precisely measured by RTK DGPS (Real-Time Kinematic Differential GPS), serve as control points and are used for data georeferencing, as well as for improving the quality of photogrammetric processing. Then, the flight for photographic acquisition can be performed. The current regulations limit the flight altitude to 150 m, but below this limit the flight altitude can be adjusted according to the study issue. Flights at lower altitudes will limit the footprint of the images and therefore the imaged area, but will improve the spatial resolution of the data. When the drone lands, the photographs are downloaded and checked. Another flight can be conducted immediately after replacing the batteries. It takes one to two hours and three operators to conduct a complete survey of an area of 50,000 m^2.

10.5. Data processing

The data are processed *a posteriori*, using dedicated software. These software solutions are based on stereophotogrammetry, a process using stereoscopy. Many software solutions are now available, among which the most widely used are PhotoScan® Pro, an easy-to-handle software developed by Agisoft©, and the open-source MicMac® tools developed over many years by IGN© (the French National Institute of Geographic and Forestry Information) [RUP 17].

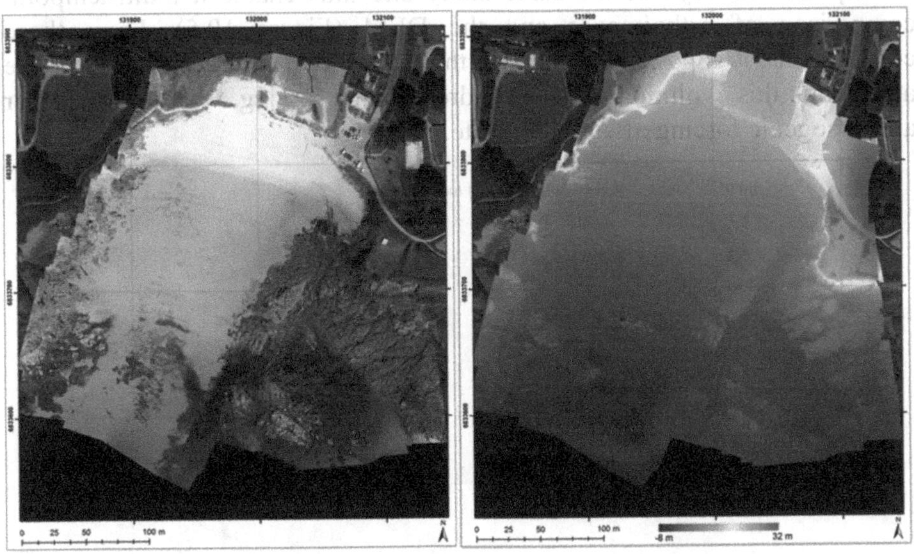

Figure 10.4. *Orthophotograph and digital elevation model (DEM) of the beach of Porsmilin (computed from 72 photographs collected in June 2015). For a color version of the figure, see www.iste.co.uk/lefloch/remote.zip*

Two types of products are generated by these photogrammetric processing systems (Figure 10.4):

– an orthophotograph or orthoimage (planimetric representation, georeferenced and geometrically corrected, which can be superimposed on a map);

– a digital elevation model (DEM), representing three-dimensional topography of the studied area.

For a flight at 100 m of altitude, the spatial resolution of the orthophoto is about 1.5 cm and the spatial resolution of the DEM is about 3 cm. The accuracy, depending on the number of targets, their spatial distribution and the quality of the GPS measurement, is of the order of 3 cm.

10.6. Examples of applications

Repeated surveys of the same study site thus enable a multi-temporal comparison of orthoimages and the DEM (Figure 10.5). From these comparisons, it is possible to deduce the morphodynamic evolution of a site and to link this evolution with the hydrodynamic forcing, weather forcing or anthropogenic forcing affecting the site.

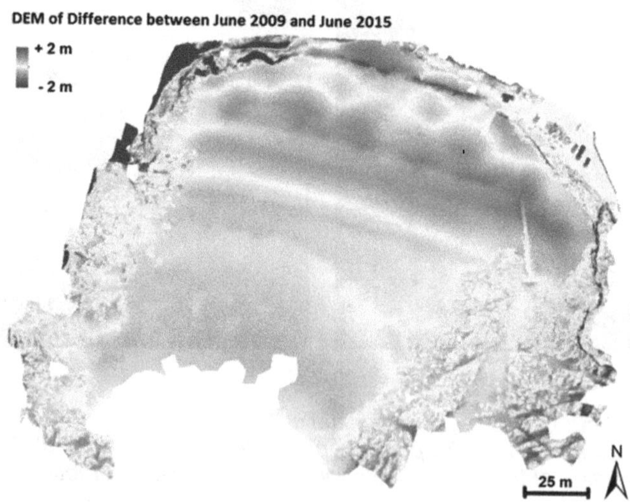

Figure 10.5. *Example of the DEM of difference made from the photogrammetric DEM representing the evolution of the beach of Porsmilin between June 2009 and June 2015. For a color version of the figure, see www.iste.co.uk/lefloch/remote.zip*

This drone has already been used in multiple coastal contexts: sandy beaches, pebble beaches (Sillon de Talbert – Pleubian), cliffs (Brittany and Normandy), estuarine mudflats at the mouth of the Seine, etc. Each coastal environment has intrinsic constraints, which sometimes require us to adapt the methodology of acquisition and/or data processing. As an example, the acquisition of nadir photographs is hardly adapted to image the vertical face of the cliffs. To solve this particular issue, the orientation of the camera is changed so as to capture the face of the cliffs with an angle of 20–30°. However, this also imposes constraints on the definition of the flight plan, which can no longer be managed by the autopilot. Similarly, mudflats are an extremely demanding environment. As moving over the site becomes extremely complex, installation and measurement of targets should be made at the minimum. Once again, by changing the acquisition strategy and flight plan, the impact of the limited number of targets can be minimized.

Figure 10.6. *Performing UAV surveys in different coastal environments: a) estuarine mudflat, b) pebble beach, c) cliff, d) sandy beach*

10.7. Conclusion

The advances in the manufacture of drones is encouraging the industry to miniaturize most of their sensors. In parallel, new UAV platforms are being developed, seeking to improve their payload capacity or autonomy in flight. Thus, new opportunities are offered to scientists and new kinds of measurements gradually complete photographic surveys: thermal measurements, Lidar, etc.

Drone potential is still increasing with the development of light and compact hyperspectral cameras, which can be onboarded on UAVs. Hyperspectral measurements allow us to gain access to water column parameters, bathymetry inversion, detection of pollution, classification of benthic habitats, etc.

10.8. References

[DEL 09] DELACOURT C., ALLEMAND P., JAUD M. *et al.*, "DRELIO: an unmanned helicopter for imaging coastal areas", *Journal of Coastal Research, Proceedings of the 10th International Coastal Symposium ICS 2009*, vol. 2, no. 56, pp. 1489–1493, 2009.

[KLE 15] KLEMAS V.V., "Coastal and environmental remote sensing from unmanned aerial vehicles: an overview", *Journal of Coastal Research*, vol. 31, no. 5, pp. 1260–1267, 2015.

[RUP 17] RUPNIK E., DAAKIR M., DESEILLIGNY M.P., "MicMac – a free, open-source solution for photogrammetry", *Open Geospatial Data, Software and Standards*, vol. 2, no. 1, pp. 1–9, 2017.

11

Use of Remote Sensing Techniques to Survey, Detect and Interpret Hydrocarbon Seeps and Spills for Exploration and Environment

11.1. Introduction

The objective of this chapter is to present remote sensing technologies and methodologies developed within Total in order to detect oil slicks and ensure spill response in the offshore domain:

– in the case of incidents occurring on surface/subsurface facilities operated by Total or nearby facilities operated by others;

– for regular monitoring of offshore production facilities.

Research concerning new emerging technologies in the remote sensing area is also discussed and several actual field cases are presented (seeps, spills, boat sewages).

In the offshore domain, the majority of satellite data used for hydrocarbon detection consists of radar (SAR – Synthetic Aperture Radar) images. Due to constraints essentially related to meteorological conditions (e.g. important cloud cover), the programming of radar data, from medium to low resolution, is preferred over optical data, especially as the existence of radar constellations offers the opportunity to acquire an image any time and

---

Chapter written by Véronique MIEGEBIELLE.

anywhere in the world. The "critical" point that still remains is the programming delay, which is the elapsed time before the collection of the first image acquired on the zone of interest. This delay is fixed by the technical specificities of each SAR satellite, orbit trajectories, the localization of the area of interest (there are more daily acquisition opportunities at the poles than at the equator) and depointing capabilities of radar systems (e.g. angles of shots).

11.2. Methodology

Synthetic aperture radar (SAR) sensors are powerful tools for detecting hydrocarbons on the sea surface due to the sensitivity of the microwave signal to the surface roughness [WIL 02, ZAT 07]. Most of the electromagnetic energy is reflected or diffused by rough surfaces. In calm sea conditions, most of the energy is scattered and the level of energy backscattered by the surface is very low. On the contrary, when the wind and sea build up, backscattered levels towards the radar become measurable. Oil slicks on the sea surface tend to weaken the capillary waves, which are the cause of sea surface roughness and therefore of the measurable backscattered signal creating black areas on the image. It is thus important to have access to met-ocean data (wind, current strength and direction) to increase the reliability of oil slick detection. Indeed, as discussed above, hydrocarbon detection requires a certain surface roughness, which is directly linked to the chop created by wind.

Oil slick signature depends on wind speed as shown in Table 11.1, bearing in mind the importance of discriminating hydrocarbons from lookalikes (algae, wrinkles, islands, etc.).

SAR image analysis is based on the visual identification of black color indices that correspond to oil slicks. The main stake is to be able to differentiate between seeps (natural origin), spills and lookalikes. Currently, the method is based on the experience and expertise of a photo-interpreter. The knowledge of wind data at instants close to the satellite image acquisition time plays an important role in oil slick detection on the sea surface. Indeed, met-ocean conditions are the cause of the roughness difference between water and oil. Oil slicks are commonly detected at wind speeds between 3 and 10 m/s. In order to get wind data as close as possible to the image acquisition time, we work with Total met-ocean specialists to get access to global databases (e.g. ECMWF), and even to *in situ* data continuously gathered at some of the facilities. In some cases, specific models are required.

Wind speed (m/s)	Oil slick signature
0–2	No backscattering of the water surface; thus, hydrocarbon detection is not possible. The term "glassy sea" is used.
2–4	No impact of the wind on oil slicks. Increased likelihood of lookalikes.
4–7	No significant impact of the wind on oil slicks. Far fewer lookalikes.
7–10	Oil slicks begin to be affected by the chop.
10–12	Only thickest oil slicks remain visible because of the dispersion of thin slicks.
>12	Energy dispersal on oil slicks, even the thick ones. Oil slick detection becomes impossible.

Table 11.1. *Responses of oil slicks according to wind speed*

11.3. Offshore facilities monitoring/mining field

The possibility of acquiring several daily images of a given area of interest in the world also allows us to set up a systematic monitoring of oil facilities. Numerous areas of interest, particularly in the Gulf of Guinea, are systematically monitored. On average, satellite SAR data is acquired twice a month. However, the number of acquisitions may increase to up to two daily acquisitions in cases of emergency, as will be discussed later.

The acquisition schedule is set for a calendar year. When an acquisition is made, we receive an alert message notifying us of the availability of the data. The image is uploaded, oceanic weather data is collected and then the analysis can begin. At the same time, the most recent description of oil installations is retrieved and added to the image.

When an oil slick is detected, the limits of its contour are drawn in order to have a vector file for each acquisition date:

– in SAR images, oil facilities (RIG, platform, FPSO, etc.) are localized by their recurring occurrence in every SAR acquisition. In the case of oil slicks caused by petroleum installations, digital files are available to inform operators about their location compared to the location of detected slicks. They allow us to determine whether the slick is linked to the considered infrastructure or not. It is then necessary to ensure that the oil leak is not recurrent in the dataset already acquired and the interpretations already achieved. The presence of pipelines at the bottom of the water should also be taken into account. These infrastructures are not visible and may also cause leaks on the ocean surface. From this point of view, having the most up-to-date facility plans is also an important stake. Monitoring helps us to locate hydrocarbon leaks and their origin, as well as to ensure their non-recurrence. Ensuring the integrity of facilities is made possible by this method of analysis;

– in the case of a degassing vessel, it is possible to link the position of the oil slick with the position of the boat, bearing in mind that the presence of the boat is not recurrent from a radar image to another. The knowledge of the SAR image acquisition date and time allows us to identify the boat thanks to its AIS signal;

– finally, in the field of exploration, in order to detect oil and gas offshore fields, we acquire a SAR database on the area of interest at different dates. We proceed in the same way as for the monitoring of facilities, by performing a manual interpretation of each image. The combination of all the analyses allows us to highlight the recurrence of hydrocarbon indices: vector files at different dates are superimposed together, and the repetitiveness of the oil slick in time is characterized by the appearance of particular structures called "flower structures". The dimensions of oil slicks vary between one and several dozen kilometers. The morphology of each "petal" of the structure is representative of a single slick. Those "petals" directly emerge from the emission point in depth, in a relatively small radius despite the depth of the water column that can reach up to 2,000 m, and then start drifting with currents on the surface. The relationship between the appearance of these "flower structures" on the surface and the morphology of the ocean floor is also a matter of interest, but that is another story which we will not develop here.

11.4. Emergency

In the framework of pollution control at Total, the coordination of three entities (remote sensing, met-ocean and environment) allowed the development of the Spill Watch tool for slick detection, monitoring and drift prediction. This tool is routinely used in cases of identified crises. Spill Watch integrates oceanographic and meteorological monitoring, modeling of winds and currents and modeling and reliable prediction of the oil slick drift up to five days after the spill.

The different stages of Spill Watch are:

– creating the met-ocean model: for several years, met-ocean specialists at Total have been working to calibrate and validate reliable models for winds and currents. In this sense, satellites continuously gathering data have proved to be rather reliable. A known hydrodynamic model is run on the acquired data to process information and allows us to obtain the direction and the intensity of currents in the water column. Models are then adjusted with data measured on-site. Total uses a system that allows us to obtain on-site data. In this way, in a five-hour time span, met-ocean models give information on currents and winds for the current day and also forecast for the next five days;

– scheduling SAR satellite acquisitions on the area of interest: we contact SAR satellite operators to know what satellite will be the first to fly over the critical area and thus program the first possible acquisition on the impacted area. As soon as an image of the area is acquired, it is transmitted to the teams for analysis, which goes through the following several processes: decoding, suppression of artifacts, georeferencing and contrast enhancement. The vector file contouring oil slicks obtained by the image analysis is then incorporated into Spill Watch. The drift modeling takes into account the actual location of the slick. The drift is adjusted and corrected at each new reception of a SAR satellite image. It allows us to significantly reduce the uncertainty on the results given by the model. In the same way and every day, all incoming data is updated to produce a new forecasting model;

– the data given by the model for D+1 is exploited by the intervention teams who work close to the slicks. They install booms to contain the slick and pumping systems to retrieve it;

– as for the data given for D+5, it allows us to anticipate coastal pollution. Specialized clean-up teams can prevent threats and deploy coastal protection means.

11.5. Perspectives

As we mentioned before, SAR remote sensing techniques are of major interest for oil slick detection in an operational context. The need for fast detection and characterization is a key factor in the HSE and exploration domains. We work in R&D on these subjects with numerous objectives, including taking into account the physical behavior of the SAR signal to detect slicks. Several parameters exploiting polarimetric information [BRE 05, SOL 12] of the electromagnetic signal backscattered by the sea surface, whether covered by hydrocarbons or not, have been proposed in the literature. We work on the hierarchical organization of these parameters based on a quantitative analysis of their detection capabilities [ANG 15].

11.6. Conclusion

The implementation of oil slick monitoring, at frequencies defined by our operational needs, provides the opportunity to:

– detect hydrocarbon traces and their recurrence (rejecting lookalikes such as algal bloom, wind and current effects, and ghost images);

– identify and classify the different possible origins of these slicks: natural seeps, spills linked to oil facilities, sewage generated by boats;

– localize the source of the spill (oil rig, pipe, underwater well, production facility, etc.);

– evaluate the magnitude of the leak, the spill and their evolution in time;

– assist the teams in charge of the facilities and the antipollution squads in mitigating the leak and treating the spill. If the leak is not marginal, the Spill Watch procedure is then deployed to follow and predict the evolution and displacement of the oil spill. Maps are then generated, which display the areas that need intervention.

This monitoring of offshore facilities provides regular surveys, information on the origin of hydrocarbon leaks (seeps or spills), near real-time surveys to help operators solve problems, maps and historical and forecast data necessary for crisis units. It has also proven to be of great use to alert other operators who are not aware of the problems in their facilities and potentially local authorities.

Given the stakes, a major research program has been initiated within Total to improve the detection, classification and quantification of oil slicks. It includes more reliable remote sensing methods, an automation of these methods, and the characterization of hydrocarbons through the simultaneous use of SAR and optical technologies.

11.7. References

[ANG 15] ANGELLIAUME S., DUBOIS-FERNANDEZ P., MIEGEBIELLE V. *et al.*, "Polarimetric parameters for oil slicks detection using SAR data remote sensing - An evaluation", *IGARSS Geoscience and Remote Sensing Symposium*, IEEE International, Milan, Italy, July 26–31, 2015.

[BRE 05] BREKKE C., SOLBERG A.H.S., "Oil spill detection by satellite remote sensing", *Remote Sensing of Environment*, vol. 95, no. 1, pp. 1–13, 2005.

[SOL 12] SOLBERG A.H.S., "Remote sensing of ocean oil-spill pollution", *Proceedings of the IEEE*, vol. 100, no. 10, pp. 2931–2945, 2012.

[WIL 02] WILLIAMS A., LAWRENCE G., "The role of satellite seep detection in exploring the south atlantic's ultradeep water", in SHUMACHER D., LESCHACK L.A. (eds), *Surface Exploration Case Histories: Applications of Geochemistry, Magnetic, and Remote Sensing*, American Association of Petroleum Geologists and the Society of Exploration Geophysicists, Tulsa, 2002.

[ZAT 07] ZATYAGALOVA V.V., IVANOV A.Y., GOLUBOV B.N., "Application of ENVISAT SAR imagery for mapping and estimation of natural oil seeps in the South Caspian Sea", *Proceedings of the ENVISAT Symposium*, Montreux, Switzerland, April 23–27, 2007.

12

Natural Escapes of Oil in Sedimentary Basins: Space-borne Recognition and Pairing with Seafloor and Sub-seafloor Features

12.1. Introduction

Natural seepage of oil occurs in most sedimentary basins worldwide, in petroliferous basins (Figure 12.1). Hydrocarbons are generated from both biogenic processes associated with burial diagenesis, which produce light gas (methane), and thermocatalytic processes, which produce heavier hydrocarbons, potentially oils.

In cases of insufficient cap rock efficiency, stocked hydrocarbons among petroleum reservoirs migrate through the sedimentary series, using fault bypass systems or stratigraphic discontinuities to be subsequently naturally released from the seafloor. It is estimated that 600,000 tons of hydrocarbons are expelled each year over the world, including biogenic and thermogenic seeps, bounded by uncertainties of 200,000 tons [WIL 73]. The release of hydrocarbon from the seafloor is typically expressed as a pockmark erosive feature, which is a conical-like seafloor depression associated with contemporary focused fluid flow [KIN 70].

Chapter written by Romain JATIAULT.

Most spectacular seeping provinces are in the Gulf of Mexico and South Caspian Sea, where hundreds of recurrent oil seeps have been revealed from the analysis of space-borne SAR imagery [ZAT 07, GAR 10].

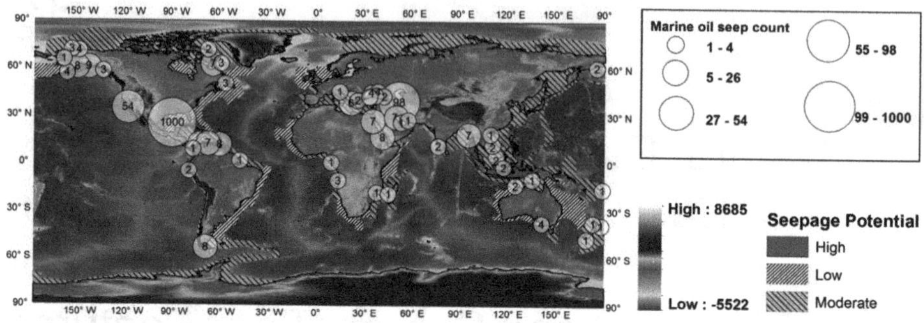

Figure 12.1. *Inventory of published locations of natural occurrence of oil seeps, superimposed with seepage potential estimated from geological context [WIL 73]. Dashed yellow: low-potential provinces; dashed orange and red: moderate- to high-potential seepage provinces. For a color version of the figure, see www.iste.co.uk/lefloch/remote.zip*

Complementing these areas, the Congo Delta is considered as an active petroleum province where geophysical investigations of the seafloor allowed the recognition of high-density fluid flow features [GAY 03, AND 11a, AND 11b] with a probable thermogenic origin [AND 12, ANK 13].

In the case of oil escapes, buoyant hydrocarbons migrate through the water column towards the sea surface, while heavier matter remains stocked on the seafloor [VAL 10, KES 11]. Depending on surface current strength, oil droplets are horizontally deflected towards their surfacing impact area at the sea surface while rising through the water column. Understanding the pathways of oil droplets is necessary to identify the seafloor origin of leaking oils.

The objectives of this presentation are to show the efficiency of SAR data to understand seepage situation in the Lower Congo Basin and to present the methodology used to identify recurrent oil slicks and the limitations of the technique. In addition, this chapter integrates measurement of underwater currents assessed from ADCP mooring points to understand the migration pathways of oil droplets and the behavior of oil plumes when rising through the water column. Finally, the integration of surface slicks, current measurements and seafloor expression of fluid flow will make it possible to pair seafloor seeps to sea surface oil slicks.

12.2. Datasets and methods

12.2.1. *Data*

12.2.1.1. *Space-borne tools to detect seeping hydrocarbons*

Synthetic aperture radar (SAR) is sensitive to capillary waves of centimetric scale induced by the local wind at the sea surface [FRA 02, MCC 03]. Oil surfactants smooth the sea surface and significantly reduce capillary waves, which triggers a contrast in roughness between oily covered areas and free water. Sea surface dampening may be induced by a large inventory of look-alike seep features, for instance bioturbation threadlike films [GAD 98, ESP 00] or low wind conditions [ESP 99]. Petroleum platforms or boat spills induce significant pitfalls [JOH 00, TRI 10, LEI 12]. Anthropogenic activities were discarded from the location of boats or platforms as they appear as shinny diffracting points in SAR data.

We used a large collection of overlapping SAR scenes in order to identify the location of recurrent oil seeps. Archive data acquired from 1994 to 2012 were used to detect recurrent oil slicks and afterward to identify the sea surface location of natural oil seepage. The dataset is mainly composed of ENVISAT ASAR (Advanced Synthetic Aperture Radar) scenes acquired between 2002 and 2012 by the European Space Agency. ENVISAT ASAR operates in C-Band with an image mode precision (IMP) and a wide swath mode (WSM). We complemented our dataset with both C-band Radarsat, ERS (1 and 2) data and X-band Cosmo-Skymmed (2009–2014) data. Once stacked, the coverage density of the SAR data reaches about 150 scenes (Figure 12.2).

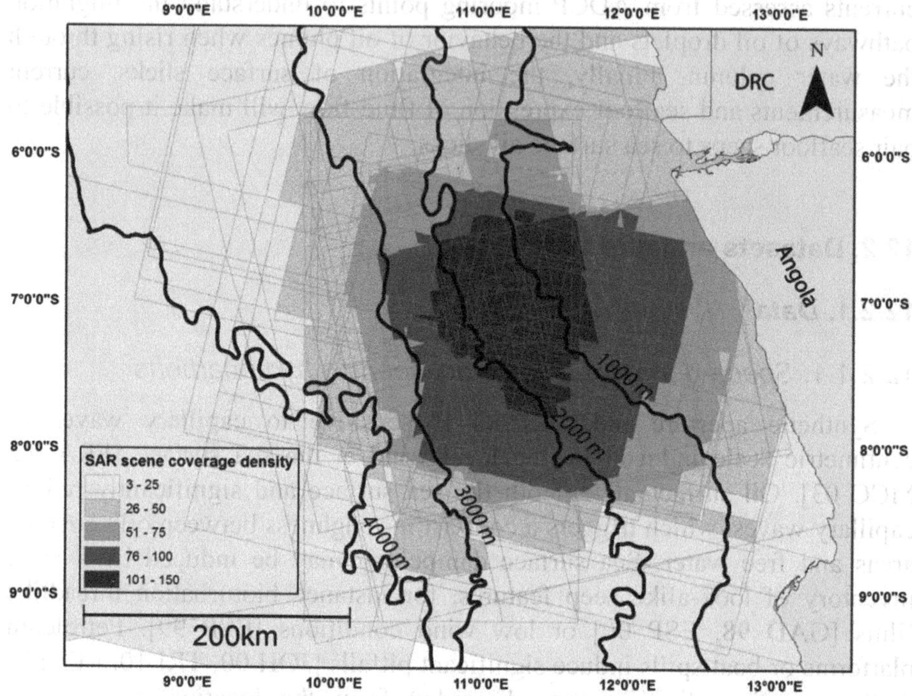

Figure 12.2. *Stack of SAR data coverage used to identify recurrent natural seepage slicks. For a color version of the figure, see www.iste.co.uk/lefloch/remote.zip*

12.2.1.2. *Assessment of current velocity across the water column*

The current velocity was measured using ADCP (Acoustic Doppler Current Profiler) tools, which use particle displacement to quantify current speed [ROW 79, GOR 96]. The data were collected over 6 months of mooring between January 2009 and July 2009 and provide measurements of current orientation and velocity along a mooring line from a series of 62 sensors across the water column. The water depth of the seafloor is about 1,800 m below sea level.

12.2.1.3. *Imaging sedimentary series using petroleum exploration data*

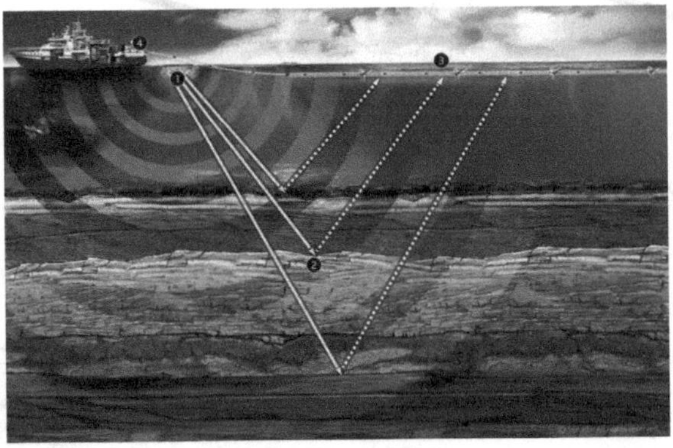

Figure 12.3. *Principle of acoustic wave propagation for imaging sedimentary series in marine environment (source: www.sercel.com; accessed 2016)*

Seismic data are widely used in petroleum companies in both onshore and offshore domains. In marine environments, air guns send acoustic waves from the sea surface that propagates through the water column and sedimentary series. Stratigraphic unconformities trigger a contrast in impedance, which reflects acoustic waves towards the sea surface. Geophysical sensors are deployed at the sea surface along streamers and record propagation times of reflected waves to estimate the vertical distance of stratigraphic reflectors (Figure 12.3). This method was developed during the 1920s by Schlumberger and is now extensively used as an exploration tool for petroleum companies, as it provides a remote imaging technique of seafloor and sub-seafloor conditions.

12.2.2. *Methods*

Because of its sensitivity to centimeter wavelengths of the surface roughness [MCC 03], synthetic aperture radar (SAR) is a remotely operated tool widely used in ocean studies for the recognition of sea surface patterns and more particularly for the detection of liquid hydrocarbons. The presence of oil surfactants dampens the sea surface and significantly reduces the amount of energy backscattered to the satellite (Fresnel specular reflection, illustration in Figure 12.4) compared to the surrounding free water. Sea surface covered by oil seepage appears as black streaks on SAR images [JOH 00, ESP 00, WIL 02, IVA 07, ZAT 07, GAR 09, KOR 14].

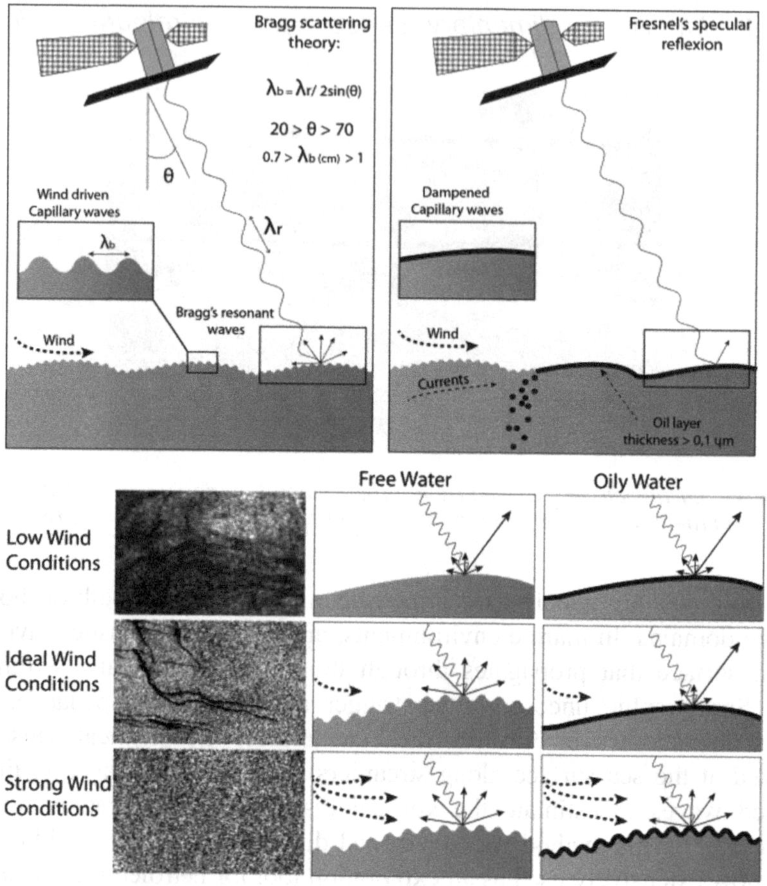

Figure 12.4. *Identification of floating oil at the sea surface using space-borne SAR*

As the sea surface, the shape of natural oil slicks is induced by the strength and orientation of the wind and surface currents [ESP 99]. Wind and surface currents induce short-term variations of the sea surface dynamics. The sinuosity of black patterns is a first-order criterion to detect oil content slicks (Figure 12.4). Sharp radiometric value contrasts between oil-covered areas and free water correspond to the sideward detectable edges of the slicks and allow us to identify the extent of individual seepage slicks (see examples in Figure 12.4).

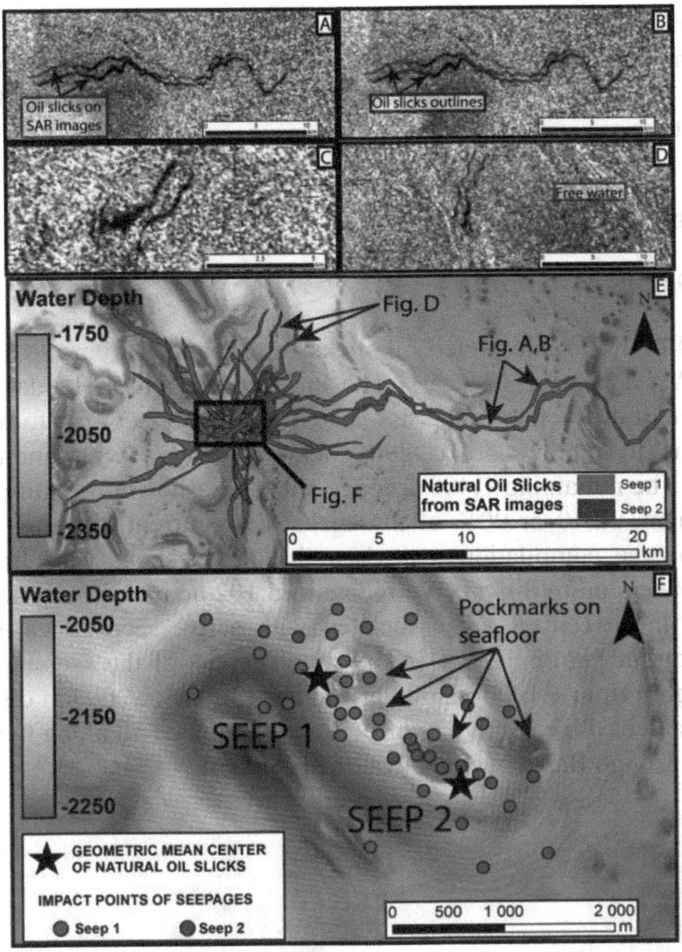

Figure 12.5. *A) Extract from ASAR WSM SAR scene dated March 26, 2012. The gray speckled background represents backscattered energy from sea surface free water. The black streaks depict the location of seepage slicks (modified from [JAT 17]). B) Same SAR image extract complemented with manual picking. C and D) ASAR WSM SAR sub-scenes dated, respectively, February 25, 2012, and December 7, 2002, superimposed with manual picking of oil slicks. The three different SAR extracts overlap at a common spatial location. The black arrows show the direction flow of surface currents during SAR scene acquisition. E) Compilation map of oil slicks outline mapped from the entire dataset, including slicks shown in A, B, C and D at two major seeping sites. All scenes are provided by the European Spatial Agency (ESA). In superposition with seafloor data measured with multibeam echosounder at 12.5 m resolution. F) Zoom at the location of seeping area revealing seafloor erosive features (pockmarks). The colored dots are oil slick origins (OSOs) [GAR 09]. The black stars are computed from the centroid of the individual OSO. For a color version of the figure, see www.iste.co.uk/lefloch/remote.zip*

Once mapped, diverging black streaks show evidence of active seeping areas of thermogenic fluids [ZAT 07, GAR 09, KOR 14]. The center of each diverging structure corresponds to the probable impact area of the fresher oil droplets at the sea surface.

12.3. Results

12.3.1. *Oil slick mapping*

A total of 1,400 interpreted oil slicks were mapped from the analysis of the 82 SAR data over 18 years. The integration of the IHS database and recognition of diffracting points associated with the boat or platform location allowed us to remove about 200 slicks associated with anthropogenic activities. The remaining 1,400 seepage slicks are shown in Figure 12.5 (A). Among them, 1,240 or 90% of the slicks are recurrent, forming diverging patterns on the compilation map. Lengths of the slicks vary from a few hundred meters up to 80 km. We recognized 102 active oil seep sites whose geometric mean centers are associated with active seep zones, shown as colored dots in Figure 12.5 (B). The water depths of the seeping province range roughly from 1,100 to 2,600 m and are 130 km to 200 km distant from the Angolan coastline. Active seep sites are distributed all along the southern part of the Congo Basin over an area of 21,000 km^2.

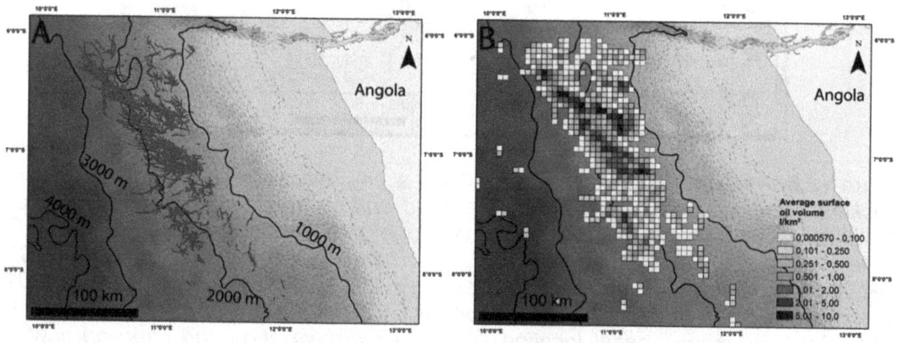

Figure 12.6. *A) Compilation map of the outline of sea surface slicks (red polygons) mapped from the analysis of SAR scene collection. B) Average volume of sea surface oil during the long-term monitoring gridded by square cells of 5 km^2. For a color version of the figure, see www.iste.co.uk/lefloch/remote.zip*

The volume of oil was computed from the amount of oil detected from the analysis of overlapping scenes normalized to the coverage density of SAR stack. The addition of each cell across the basin suggests that an average of 1,625 liters of oil are currently surfacing on the ocean surface. This volume was computed considering a random slick thickness of 0.145 um, estimated from *in situ* measurements in the Gulf of Mexico and based on the Bonn Agreement. We estimated that the amount of oil expelled from natural seeps is about 4,400 m^3 each year.

12.3.2. *Oil migration pathways and horizontal deflection*

Oil droplets are horizontally deflected from their seafloor source point towards the surfacing area while rising through the water column. The deflection value is controlled by underwater current strength. In order to quantify the horizontal deflection during SAR acquisition, we order SAR scenes during ADCP measurements.

The integration of ADCP data allowed us to compute interval values. The summation in direction and distance between each interval allowed us to quantify the horizontal deflection during the SAR acquisition.

The mooring area is located 15 km westward from a major seeping area recognized for the interpretation of SAR collection. We considered the location of the sea surface impact area of the seepage slick, namely the oil slick origin [GAR 09], to preserve deflection within the water column freely from sea surface drift. The slicks detected on June 30, 2009 (Figure 12.7) lie over an area previously recognized for a high seepage activity (Seep 1 in Figure 12.5 (F)).

Figure 12.7 (B) shows that the horizontal deflection value remains low compared to the water depth. Considering a uniform rising speed across the water column, a slow rising speed of 2 cm/s produces a deflection value below 1,500 m.

The seafloor origin is well identifiable from seafloor and sub-seafloor data. The analysis of seafloor features reveals conical-like pockmarks typical of natural fluid seeps (Figure 12.7 (C)).

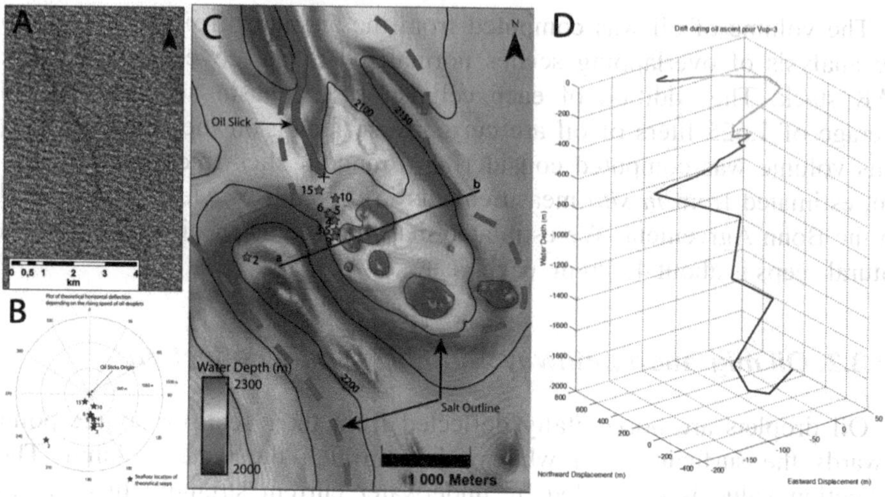

Figure 12.7. *A) Extract of SAR scene dated from June 30, 2009, allowing the recognition of oil surface-covered areas (black patches). The green points are two centroids of OSOs. B) Backward modeling of theoretical deflection (m) from sea surface OSO to seafloor seep. Computed for different rising speeds between 2 and 15 cm/s (labeled next to red stars) by integrating ADCP data measurements (15 km from seeping area). C) Seafloor map (isochron superimposed with dip map) in addition to the oil slick location (Figure 12.5 (F)), salt outline (extracted from 3D seismic) and backward deflections with different rising speeds. Best fit between seafloor pockmark location and sea surface slicks is obtained for a rising speed of 3 cm/s across the water column. D) Theoretical migration pathway of oil droplet through the water column from seafloor source to the seepage slicks surfacing point (rising speed of 3 cm/s) (modified after [JAT 18]). For a color version of the figure, see www.iste.co.uk/lefloch/remote.zip*

The analysis of underlying geophysical profiles shows that the active pockmark is located at the rim of evaporitic diapirs (mainly composed of Halite, Gypsum, etc.), which is known to create efficient migration pathways for migrating oils towards the seafloor in the area, at the sedimentary/evaporitic interface [GAY 03, AND 12, SER 17]. In addition, hydrocarbon seeps are usually associated with high-impedance patches close to the seafloor. Anaerobic oxidation of methane creates hardground authigenic carbonates [HOV 87], which are reflective series on seismic sections and typical of hydrocarbon seepage systems. Figure 12.8 shows the location of salt diapir, location of seafloor pockmark and migration chimney made of carbonates. These characteristics suggest that the pockmark is obviously hydrocarbon-active and the origin of expelled oil.

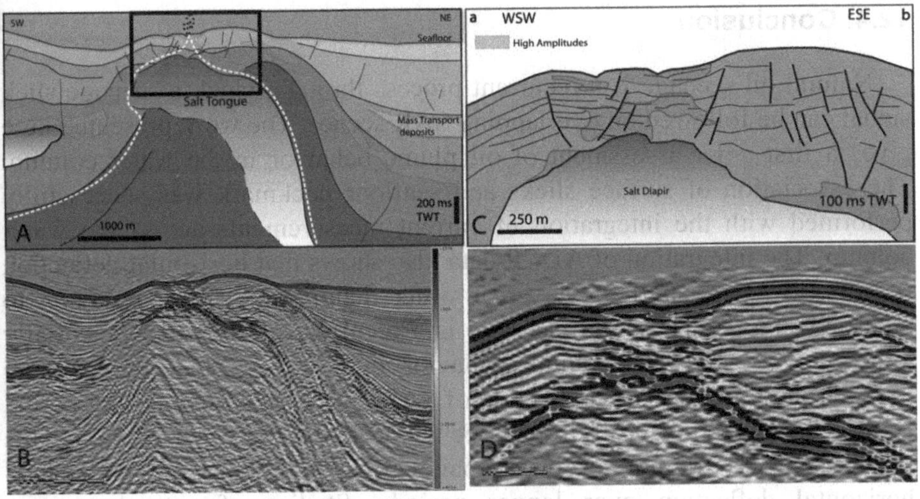

Figure 12.8. *A, B) line drawing and wide seismic cross section through underlying sedimentary series at the location of the seeping site. The white dashed line represents the probable migration pathways. C, D) line drawing and seismic section focus at shallower depths allowing the recognition of high amplitudes on seismic sections, fault network, salt diapirs. For a color version of the figure, see www.iste.co.uk/lefloch/remote.zip*

The theoretical deflection value computed for high rising speeds of 15 cm/s projected on the seafloor is located far from any seafloor seep manifestations. The progressive decrease of rising speeds brings the theoretical location closer to the seafloor seeps (Figure 12.7 (C)). The best correlation is between observed and computed locations of fluid source gas for a rising speed of 3 cm/s, the migration pathway of which is shown in Figure 12.8 (D).

The modeled migration profile suggests distinctive water mass units with reverse directions. The horizontal deflection resulting from the addition of opposite currents is expressed through a low horizontal deflection.

Garcia-Pineda *et al.* [GAR 09] established a linear relationship between the water column and horizontal deflection offset of oil droplets in the Gulf of Mexico. The hydrodynamics of the Lower Congo Basin might be inflected by a reverse current component, which produces a restricted horizontal deflection and surfacing area of oil droplets.

12.4. Conclusion

Natural oil seepage is a recurrent process through which a seepage slick signal can be followed on overlapping SAR scenes. The work presented here gives a first-order assessment of oil plume behavior in the water column. The association of surface slicks and seafloor pockmark was successfully performed with the integration of current measurements across the water column. The integration of ADCP data also shows that horizontal deflection remains low compared to the great water depth value. In comparison to anthropogenic spills, natural seepage is a recurrent process that potentially allows us to study the plume behavior over a consistent period. The integration of an extended collection of SAR data during the mooring period is necessary to validate oil droplets behavior across the water column. This should help us to monitor oil seepage affected regions and understand horizontal deflection over longer periods. Studies of natural seepage complement the evaluation of oil/gas plume behavior from accidental spills and could also help in emergency response.

12.5. References

[AND 11a] ANDRESEN K.J., HUUSE M., SHØDT N.H. *et al.*, "Hydrocarbon plumbing systems of salt minibasins offshore Angola revealed by three-dimensional seismic analysis", *AAPG Bulletin*, vol. 95, no. 6, pp. 1039–1065, 2011.

[AND 11b] ANDRESEN K.J., HUUSE M., "'Bulls-eye' and polygonal faulting in the Lower Congo Basin: relative timing and implications for fluid expulsion during shallow burial", *Marine Geology*, vol. 279, nos 1–4, pp. 111–127, 2011.

[AND 12] ANDRESEN K.J., "Fluid flow features in hydrocarbon plumbing systems: what do they tell us about the basin evolution", *Marine Geology*, vols 332–334, pp. 89–108, 2012.

[ANK 13] ANKA Z., ONDRAK R., KOWITZ A. *et al.*, "Identification and numerical modeling of hydrocarbon leakage in the Lower Congo Basin: implications on the genesis ok km-wide seafloor mounded structures", *Tectonophysics*, vol. 604, pp. 153–171, 2013.

[ESP 99] ESPEDAL H.A., WAHL T., "Satellite SAR oil spill detection using wind history information", *International Journal of Remote Sensing*, vol. 20, no. 1, pp. 49–65, 1999.

[ESP 00] ESPEDAL H.A., JOHANNESSEN O.M., "Detection of oil spills near offshore installations using synthetic aperture radar (SAR)", *International Journal of Remote Sensing*, vol. 21, no. 11, pp. 2141–2144, 2000.

[FRA 02] FRANCESCHETTI G., IODICE A., RICCIO D. *et al.*, "SAR raw signal simulation of oil slicks in ocean environments", *IEEE Transactions on Geoscience and Remote Sensing*, vol. 40, no. 9, pp. 1935–1949, 2002.

[GAD 98] GADE M., ALPERS W., HÜHNERFUSS H. *et al.*, "Imaging of biogenic and anthropogenic ocean surface films by the multifrequency/multipolarization SIR-C/X-SAR", *Journal of Geophysical Research: Oceans*, vol. 103, no. C9, pp. 18 851–18 866, 1998.

[GAR 09] GARCIA-PINEDA O., ZIMMER B., HOWARD M. *et al.*, "Using SAR images to delineate ocean oil slicks with a texture-classifying neural network algorithm (TCNNA)", *Canadian Journal of Remote Sensing*, vol. 35, no. 5, pp. 411–421, 2009.

[GAR 10] GARCIA-PINEDA O., MACDONALD I., ZIMMER B. *et al.*, "Remote-sensing evaluation of geophysical anomaly sites in the outer continental slope, northern Gulf of Mexico", *Deep Sea Research Part II: Topical Studies in Oceanography*, vol. 57, nos 21–23, pp. 1859–1869, 2010.

[GAY 03] GAY A., LOPEZ M., COCHONAT P. *et al.*, "Sinuous pockmark belt as indicator of a shallow buried turbiditic channel on the lower slope of the Congo Basin, West African Margin", in VAN RENSBERGEN P., HILLIS R.R., MALTMAN A.J., MORLEY C.K. (eds), *Subsurface Sediment Mobilization,* Geological Society of London, London, 2003.

[GOR 96] GORDON R.L., Acoustic Doppler current profiler principles of operation: a practical primer, Teledyne RD Instruments, San Diego, 1996.

[HOV 87] HOVLAND M., TALBOT M.R., QVALE H. *et al.*, "Methane-related carbonate cements in pockmarks of the North Sea", *Journal of Sedimentary Petrology*, vol. 57, no. 5, pp. 881–892, 1987.

[IVA 07] IVANOV A.Y., GOBULOV B.N., ZATYAGALOVA B.B., "On oil and gas seeps and underground fluid discharges in the Southern Caspian based on Space Radar Data", *Earth Exploration from Space*, vol. 2, pp. 62–81, 2007.

[JAT 17] JATIAULT R., DHONT D., LONCKE L. *et al.*, "Monitoring of natural oil seepage in the Lower Congo Basin using SAR observations", *Remote Sensing of Environment*, vol. 191, pp. 258–272, 2017.

[JAT 18] JATIAULT R., DHONT D., LONCKE L. *et al.*, "Deflection of natural oil droplets through the water column in deep-water environments: the case of the Lower Congo Basin", *Deep Sea Research Part I: Oceanographic Research Papers*, vol. 136, pp. 44–61, 2018.

[JOH 00] JOHANNESSEN O.M., SANDVEN M., JENKINS A.D. *et al.*, "Satellite earth observation in operational oceanography", *Coastal Engineering*, vol. 41, nos 1–3, pp. 155–176, 2000.

[KES 11] KESSLER J.D., VALENTINE D.L., REDMOND M.C. *et al.*, "A persistent oxygen anomaly reveals the fate of spilled methane in the deep Gulf of Mexico", *Science*, vol. 331, no. 6015, pp. 312–315, 2011.

[KIN 70] KING L.H., MACLEAN B., "Pockmark on the Scotian Shelf", *Geological Society of America Bulletin*, vol. 81, no. 10, pp. 3141–3148, 1970.

[KOR 14] KÖRBER J.H., SAHLING H., PAPE T. *et al.*, "Natural oil seepage at Kobuleti Ridge, eastern Black Sea", *Marine and Petroleum Geology*, vol. 50, pp. 68–82, 2014.

[LEI 12] LEIFER I., LEHR W., SIMECEK-BEATTY D. *et al.*, "State of the art satellite and airborne marine oil spill remote sensing: application to the BP Deepwater Horizon oil spill", *Remote Sensing of Environment*, vol. 124, pp. 185–209, 2012.

[MCC 03] MCCANDLESS S.W., JACKSON C.R., Synthetic Aperture Radar Marine User's Manual, NOAA/NESDIS, 2003.

[ROW 79] ROWE F., YOUNG J., "An ocean current profiler using Doppler sonar", *Oceans '79 IEEE Proceedings*, San Diego, USA, pp. 292–297, 1979.

[SER 17] SERIÉ C., HUUSE M., SCHØDT N.H. *et al.*, "Subsurface fluid flow in the deep-water Kwanza Basin, offshore Angola", *Basin Research*, vol. 29, no. 2, pp. 149–179, 2017.

[TRI 10] TRIVERO P., BIAMINO W., "Observing marine pollution with Synthetic Aperture Radar", in IMPERATORE P., RICCIO D. (eds), *Geosciences and Remote Sensing, New Achievements*, InTech Open, London, 2010.

[VAL 10] VALENTINE D.L., KESSLER J.D., REDMOND M.C. *et al.*, "Propane respiration jump-starts microbial response to a deep oil spill", *Science*, vol. 330, no. 6001, pp. 208–211, 2010.

[WIL 73] WILSON R.D., MONAGHAN P.H., OSANIK A. *et al.*, "Estimate of annual input of petroleum to the marine environment from natural marine seepage", *Transactions of the Gulf Coast Association of Geological Societies*, vol. 23, pp. 182–193, 1973.

[WIL 02] WILLIAMS A., LAWRENCE G., "The role of satellite seep detection in exploring the south atlantic's ultradeep water", in SHUMACHER D., LESCHACK L.A. (eds), *Surface Exploration Case Histories: Applications of Geochemistry, Magnetic, and Remote Sensing*, American Association of Petroleum Geologists, Tulsa, 2002.

[ZAT 07] ZATYAGALOVA V.V., IVANOV A.Y., GOLUBOV B.N., "Application of ENVISAT SAR imagery for mapping and estimation of natural oil seeps in the South Caspian Sea", *Proceedings of the ENVISAT Symposium*, Montreux, Switzerland, April 23–27, 2007.

Conclusion

Principal Conclusions of Debates: Synthesis and Perspectives

C.1. Challenges for innovation

Tackling a rare event is a difficult challenge. Let us take the example of pollution having subsurface effects. The main platforms used for marine pollution observation are airplanes and satellites but dedicated to the sea surface. If we have to investigate the subsea surface domain, we must identify submarine platforms for operational observation. Numerous sensors already exist and promise potential solutions. However, chemical pollutions are rare [MAM 09], suggesting weak incitation in terms of investment. The development of these solutions may suffer from a lack of interest from companies and laboratories, as well as financial support from institutions.

Addressing a popular innovation may be a risky approach. Unmanned aerial vehicles (UAVs), commonly known as drones, have become must-have products. This is a consequence of many factors such as their ubiquity in daily life, their acceptance by authorities and European industrial priorities [SES 18]. Numerous drone technologies could be of interest as well as the small platforms already presented. Nevertheless, a major drawback can intervene with any promising solution in terms of reliability

---

Conclusion written by Stéphane LE FLOCH and Frédéric MUTTIN.

and maturity. A lack of concern among others should directly attribute funds and practices to only one technology.

The complementary debate about the popularity of one solution is the difficulty of disposing of several tools and making the adequate choice. In fact, any technology has advantages and drawbacks. Consequently, the methodology suggested is an integrative approach involving a selection between different tools. First, data sources should be as exhaustive as possible. Second, a classification can deliver an optimal choice based on a particular operational objective. We consider drones as complementary solutions to airplanes and satellites. Flexibility and availability are two advantages of drones. Nevertheless, compared to past decades when data sources were scare, today, the situation is reversed. The approach is now to face multiple sources of data and choose the most accurate one. Decision aid systems could help manage numerous possibilities.

A promising research result can deliver benefits to existing methods. Let us address the perspectives of hyperspectral imagery [BON 09, SOL 12]. This technology can identify the size of water bubbles and the emulsification ratio inside an oil layer. These parameters should be included in oil-weathering models, themselves embedded in an oil-drifting model. More precisely, the weathering process embedded in a model nowadays presents oil as fresh, or sometimes with an aging limited to some hours or to one or two days. The future approach will measure the water proportion inside the oil in real time. Consequently, an oil viscosity update could follow in real time. We propose an investigation into forecasts of oil spreading and slick geometry, taking into account hyperspectral imagery. The authorities ask for better predictions of oil slick weathering and displacement processes, with simulations spanning three to five days. We underline that oil-drifting models have become more and more complex.

C.2. Preliminary juridical aspects

A further discussion looks into oil discharge from ships as having more mature juridical aspects than for chemicals. The aim of this debate is to describe former experiences, bringing to the audience the lessons learned and guidance for best practices.

Teledetection performs insufficiently to formally recognize a product even with the compilation of different sensor results. Nevertheless, teledetection can be of help to the juridical court and is *a priori* allowed. In any case, teledetection by itself is insufficient as the judgment base.

Images of pollution can give valuable information with a judicial value. When added with other clues, the deliberate aspect of the ship discharge can be proved. We take as example a pollution case detected by satellite imagery in the English Channel–North Sea outside the territorial zone. We have the proof of the deliberate aspect of this discharge by coupling expert consultations and ship crew questionings.

In fact, a pollutant identification must be made by a judicial inquest. The investigation includes an analysis of the hydrocarbon register of the ship in the application of annex 1 of the MARPOL convention [MAR 73]. An inspection of the ship is also carried out to formally identify all hydrocarbons stored onboard.

Ships dispose of an oil–water separator to allow fluid discharges of at least 15 parts per million of hydrocarbons [MAR 73]. In fact, the international legislation permits such infinitesimal pollution, understanding that large efforts have been made in this field. The threshold was 100 parts per million several decades ago. The onboard separator device allows the checking of the oil concentration before discharge. A difficult choice arises for the crew concerning the practical usage of the separator, particularly when the concentration computed by the analyzer reaches the avoidance limit.

For the identification of perpetrators, the adding of tracers to the hydrocarbons has been suggested for several years. By identifying the tracer, the hydrocarbon composition, the ship name can be deduced. The first difficulty is to identify the different parties knowing the data, the properties of the tracers, the tracer inserted in a particular product and the identity of the ship carrying this marked hydrocarbon. A further issue concerns the access to these data for the defender during judicial processes. The risk is highlighted in both technical and judicial debates. The primary point of this method is to focus on a particular ship identification as opposed to the hydrocarbon kind of identification.

Demonstrating a physical proof during a court audience is a major challenge with a worldwide dimension. We provide non-exhaustive particular cases that are highly difficult to solve. A sample of gaseous discharge should be made safely and before spreading. A sea surface image suspected of showing interferences, instead of pollution, could agree with the other images of classical interferences on sea, without pollution and similar water surface capture. We suggest the use of underwater drones for identifying pollution dissolved in the water column. We expect judicial usage of acoustic waves from multibeam sounders.

C.3. Interference of physical factors on sea

Images can superpose the effects of several physical factors on the sea surface that renders the task of pollution detection difficult. We provide non-exhaustive sources of interferences, some companies and institutes studying this topic and common drawbacks.

In satellite images, the initially cited interferences in marine pollution detection are the sea ice coverage and surface roughness with wind dependence. The interferences can affect the detected boundary of the pollution pattern. We can also mention the presence of suspended matter in the seawater and the effect of the sea surface temperature. Another source of interference is the presence of several pollutions mixing different responses for each of them. The concentration of floating wasted materials gives a modification of the sea surface characteristics. The accumulation of algae in rafts allows the wastes to agglomerate during drifting. The SG-Mer administration performed an analysis of macro-waste several years ago. That study revealed a high dispersion of waste. It is unlikely that floating waste agglomerates in slicks or rafts. This reduces the possibility of their tracking.

We distinguish with difficulty the macro-structure of waste and the floating sargassum and golden tide. The presence of floating waste is the main problem facing both aerial survey and teledetection.

The images from the IFREMER institute are available daily over 24 hours. The Copernicus project connects images with model results. The monitoring of pollution events uses teledetection. EMSA agency uses radar-based observations from the CleanSeaNet process. The model computes the patch contour for detected pollution. The number of false-positives reduces drastically by taking into account the sea surface temperature and the quantity of suspended matters. The company CLS, in conjunction with EMSA, computes the images in 20–30 minutes with a treatment embedding the sea surface temperature. However, a higher delay is mandatory for an image acquired over 24 hours. A research project involving the ONERA institute and Total studies the correlation between images with different origins.

C.4. Drone competitiveness

Floating drones are maritime autonomous vehicles with different sizes. The IFREMER institute develops two floating drones, one similar to a kayak and another smaller one. This latter one can perform multi-mission measurement or sampling with reduced onboard equipment for weight reduction. This permits sampling in rough seas.

The company ACTIMAR develops underwater gliders that allow long duration missions, one to two months. The measurements taken include hydrocarbon concentration in the water column. Another usage of glider sensors is the updating of the data during model running. The parameters are the water temperature and salinity. Forecast model accuracy requires data updating.

Photogrammetry is a technique that uses different 2D views of the same body, providing an image in 3D. Geographic mapping is already available to produce 3D representation, more evolved than the planar map. Mapping in 3D is a former issue. Today, the task remains to produce 3D maps showing *in situ* data and model results. Processing a useful view using an aerial drone requires precise knowledge of its position. Progress has been made concerning the geographical position control of the drone.

Stationary flights are an underused capacity of drones. Remaining at a fixed point, the drone can monitor a variable in real time. The values observed can be included in a model. We call this process data assimilation. The first usage appears in drifting slick models.

Ongoing issues are numerous. The autonomy duration of a drone must be longer. For defense purposes, the drone monitors the surveillance of an inland position for several hours or days. It is a challenge to adapt them to marine constraints. Particularly, the costs must be largely temperate without reducing the large quantity of data they can measure. The models follow the same tendency with regard to the extension of data.

C.5. Pollutant identification

Discussions address gas leakage and dissolved pollution in the water column. We note the accomplishment of tests on aerial gas emission in the Mediterranean Sea, but without access to the campaign results. We suppose that the sensitivity of sensors was insufficient. On the ship itself, we could directly observe gaseous emission. We suggest studying the onboard sensor.

Regarding NOx and SOx, the detection and quantification seem to be difficult. We plan future investigations into the visibility of these gases. A proposal to ANR will allow a follow-up to the POLLUPROOF project. The emphasis of this project focuses on the lack of the technical implementation of annexes 2 and 6 of the MARPOL convention.

There is an ongoing project undertaken by the EMSA agency to monitor SOx, NOx and CO_2 emissions using drones. In 2017, its starting year, this project permitted the allocation of three drones to the EU member states for gas detection.

The methods described for floating and aerial pollutions seem to be inaccurate to analyze a product dissolved in the water column. Over the last 10-year period, research into this topic has focused on multibeam echo sounders [STA 19, HOL 17]. This technique achieved essentially bathymetric analysis. More advanced treatment of these signals should identify suspended bubbles and droplets at specific water column depths. The available pollutant adsorption model could provide the sub-surface contaminated zone. In other words, bathymetric measurements allow investigation of pollutant fate. Nowadays, we can observe some research in this topic presented by the ONERA institute.

The analysis performed in the ONERA laboratory depicted vegetable oil fate, showing that this pollutant evolved over time with a dissolved fraction. This issue differs from the chemical description of the pollutant at the water surface. The study of several kinds of vegetable oils reveals persistency or tendency to dissolve. This research is just a beginning.

C.6. Sensors

The study of sensors can use two complementary methods. During the POLLUPROOF project, we observed the responses of chosen pollutants on radar and passive sensors. The opposite scheme was also valuable. We tried to detect and identify an unknown product using different sensors, including laser fluorosensor [SOL 12].

The first study of sampling hazardous and noxious substances concerned the safety of operators. Selected floating drones monitored seawater samples using dedicated HNS sensors. Drifting and sampling of HNS are two issues for future research in numerical modeling.

We addressed the onboard cameras experimented on by the French Navy. We emphasized the capacity of 3D cameras. Automatic cartography is incompatible with the marine environment. Images use different wavelengths and give their GPS coordinates.

A camera includes software. The public can access general image software and hence would need to accommodate maritime observation. Future software could deliver flight plans for mapping out polluted zones during morning and evening, showing the evolution.

Multibeam sounders emit acoustic waves. The name of the periodic beeped signal is “full wave”. The design of the sounder waveguide allows responses with a valuable time dimension. Recording and analyzing the full return signals will theoretically describe the parts of subsea properties. Generally, the sounder manufacturers hesitate to give their definition of “full wave”, limiting the ability to describe the media spanned.

An assemblage of a set of sensors could detect the most common pollutants. The concept of a general platform for slick identification combines hyperspectral imagery, acoustic waves and bathymetric Lidar [WAG 00, TUR 13]. This last sensor development is developing rapidly.

As an acoustic sounder, lidar emits laser light pulses traveling the water column and reflecting on the seabed. Similarly, bathymetric lidar uses a continuous recording of the return signal. The echo contains the water column effects, which include pollutant thickness and spatial dependency. Nowadays, this sensor is used for mapping the seabed, including the deep-sea region.

We expect future developments in subsea autonomous vehicles performing *in situ* measurements. Multi-spectral imagery has raised great hopes for pollution mapping. The combination of acoustic waves, Lidar, passive sensors and fluorosensors allows powerful substrate characterization.

C.7. References

[BON 09] BONIN G., DUBOIS-FERNANDEZ P., RUAULT-DU-PLESSIS O. *et al.*, "The new ONERA multispectral airborne SAR system in 2009", *Radar Conference*, IEEE International, Pasadena, USA, May 4–8, 2009.

[HOL 17] HOLLER P., MARKERT E., BARTHOLOMÄ A. *et al.*, "Tools to evaluate seafloor integrity: comparison of multi-device acoustic seafloor classifications for benthic macrofauna-driven patterns in the German Bight, southern North Sea", *Geo-Marine Letters*, vol. 37, pp. 93–109, 2017.

[MAM 09] MAMACA E., GIRIN M., LE FLOCH S. *et al.*, "Are HNS spills more dangerous than oil spills?", *Interspill Conference and the 4th IMO R&D Forum*, Marseille, France, May 12–14, 2009.

[MAR 73] MARPOL, International convention for the prevention of pollution from ships, 1973, as modified by the protocol of 1978 relating thereto and by the Protocol of 1997, International Maritime Organization, available at: http://www.imo.org (accessed June 2020), 1973.

[SES 18] SESAR JOINT UNDERTAKING, European ATM Master Plan: roadmap for the safe integration of drones into all classes of airspace, SESAR Joint Undertaking, Brussels, available at: https://www.sesarju.eu/node/2993 (accessed June 2020), 2018.

[SOL 12] SOLBERG A.H.S., "Remote sensing of ocean oil-spill pollution", *Proceedings of the IEEE*, vol. 100, no. 10, pp. 2931–2945, 2012.

[STA 19] STATECZNY A., BURDZIAKOWSKI P., "Universal autonomous control and management system for multipurpose unmanned surface vessel", *Polish Maritime Research*, vol. 26, pp. 30–39, 2019.

[TUR 13] TURCOTTE C.S., PUCKRIN E., AUBÉ F. *et al.*, "Novel compact airborne platform for remote sensing applications using the Hyper-Cam infrared", in HENRY D.J. (ed.), *Airborne Intelligence, Surveillance, Reconnaissance (ISR) Systems and Applications X,* Baltimore, United States, 2013.

[WAG 00] WAGNER P., HENGSTERMANN T., ZIELINSKI O., "MEDUSA: an airborne multispectral oil spill detection and characterization system", *Infrared Technology and Applications XXVI,* vol. 4130, pp. 610–620, 2000.

List of Authors

Pascal ALLEMAND
LGLTPE CNRS
Claude Bernard University Lyon 1
France

Jérôme AMMANN
LDO CNRS
University of Western Brittany
Plouzané
France

Sébastien ANGELLIAUME
ONERA
Salon de Provence
France

Louis ARMSTRONG
Transport Canada
Ottawa
Canada

Anne-Pier BERNIER
AEREX Avionics
Quebec
Canada

Laurent BUIGNET
French Customs
Lann-Bihoué
France

Sophie CHATAING
CEDRE
Brest
France

Lucie COCQUEMPOT
LDO CNRS
University of Western Brittany
Plouzané
France

Christophe DELACOURT
LDO CNRS
University of Western Brittany
Plouzané
France

Dennis DURO
AEREX Avionics
Quebec
Canada

Pierre-Yves FOUCHER
ONERA
Toulouse
France

Guillaume GAGNE
AEREX Avionics
Quebec
Canada

Francis GOHIN
IFREMER
Plouzané
France

Philippe GRANDJEAN
LGLTPE CNRS
Claude Bernard University Lyon 1
France

Guillaume HAJDUCH
CLS Collecte Localisation Satellites
Plouzané
France

Romain JATIAULT
Total
Pau
France

Marion JAUD
LDO CNRS
University of Western Brittany
Plouzané
France

Nicolas LE DANTEC
LDO CNRS CEREMA
Plouzané
France

Stéphane LE FLOCH
CEDRE
Brest
France

Sébastien LEGRAND
Marine Forecasting Centre
RBINS
Brussels
Belgium

Philippe MARTINEAU
ONERA
Salon de Provence
France

Véronique MIEGEBIELLE
Total
Pau
France

Brent MINCHEW
British Antarctic Survey (BAS)
Cambridge
United Kingdom

Frédéric MUTTIN
EIGSI
La Rochelle
France

Eldon PUCKRIN
DRDC Valcartier Research Centre
Quebec
Canada

Yann RABUTEAU
ALLEGANS
University of Western Brittany
Brest
France

Ronny SCHALLIER
Aerial Surveillance Team RBINS
Brussels
Belgium

Charles Henri THOUAILLE
CEPPOL French Navy
Brest
France

Index

A, B, C

D, E, F

G, H, I

S, T

U, V, W, X

Other titles from

ISTE

in

Ecological Science

2020

BRUSLÉ Jacques, QUIGNARD Jean-Pierre
Fish Behavior 1: Eco-ethology
Fish Behavior 2: Ethophysiology

ROSSIGNOL Jean-Yves
Climatic Impact of Activities: Methodological Guide for Analysis and Action

2019

AMIARD Jean-Claude
Industrial and Medical Nuclear Accidents: Environmental, Ecological, Health and Socio-economic Consequences
(Radioactive Risk SET – Volume 2)
Nuclear Accidents: Prevention and Management of an Accidental Crisis
(Radioactive Risk SET – Volume 3)

BOULEAU Gabrielle
Politicization of Ecological Issues: From Environmental Forms to Environmental Motives

GIRAULT Yves
UNESCO Global Geoparks: Tension Between Territorial Development and Heritage Enhancement

KARA Mohamed Hichem, QUIGNARD Jean-Pierre
Fishes in Lagoons and Estuaries in the Mediterranean 2: Sedentary Fish
Fishes in Lagoons and Estuaries in the Mediterranean 3A: Migratory Fish
Fishes in Lagoons and Estuaries in the Mediterranean 3B: Migratory Fish

OUVRARD Benjamin, STENGER Anne
Incentives and Environmental Policies: From Theory to Empirical Novelties

2018

AMIARD Jean-Claude
Military Nuclear Accidents: Environmental, Ecological, Health and Socio-economic Consequences
(Radioactive Risk SET – Volume 1)

FLIPO Fabrice
The Coming Authoritarian Ecology

GUILLOUX Bleuenn
Marine Genetic Resources, R&D and the Law 1: Complex Objects of Use

KARA Mohamed Hichem, QUIGNARD Jean-Pierre
Fishes in Lagoons and Estuaries in the Mediterranean 1: Diversity, Bioecology and Exploitation

2016

BAGNÈRES Anne-Geneviève, HOSSAERT-MCKEY Martine
Chemical Ecology

2014

DE LARMINAT Philippe
Climate Change: Identification and Projections